MANUAL OF
ELECTRONIC SERVICING
TESTS AND MEASUREMENTS

Robert C. Genn, Jr.

Illustrated by E. L. Genn

PARKER PUBLISHING COMPANY, INC.

WEST NYACK, NEW YORK

Other Books by the Author:

Workbench Guide to Electronic Troubleshooting

Practical Handbook of Low-Cost Electronic Test Equipment

©1980 *by*

Parker Publishing Company, Inc.
West Nyack, New York

Library of Congress Cataloging in Publication Data

Genn, Robert C
    Manual of electronic servicing tests and measurements.

    Includes index.
    1. Electronic apparatus and appliances--Testing.
2. Electronic apparatus and appliances--Maintenance and
repair. 3. Electronic measurements. I. Genn, E. L.
II. Title.
TK7870.G4218   621.381'028   79-9417
ISBN 0-13-553388-0

Printed in the United States of America

How This Book Will Help You Perform
Virtually Any Electronic Testing Job

If you've been looking for a ready reference covering every aspect of electronics testing, here it is—*a modern source book* of test techniques that will enable you to solve quickly virtually any electronics testing problem found in the average service shop. You'll learn how to make and analyze the results of *one-hundred and ninety-seven tests and measurements* as well as how to come to the correct conclusions that are absolutely essential for successful troubleshooting. Each chapter takes you step-by-step through simplified testing procedures that are of real, practical value to technicians and experimenters.

In Chapter One semiconductor tests and measurements are analyzed in detail. An explanation of how to test modern FET's and dozens of other contemporary solid state components is included. The second chapter contains an analysis of all common radio receiver tests and how to make them—everything from the RF mixer stages to the speaker system. Both solid state and tube-type receivers are covered. Next, comprehensive guidelines on audio equipment tests and measurements are given in Chapter Three. Again, you'll find both solid state and tube-type equipment.

Chapter Four includes a wide range of practical testing ideas, techniques and shortcuts based on actual shop experience. They will be helpful regardless of your primary interest—audio or RF, and each one will improve your ability to use test instruments skillfully and easily for fast, accurate results.

Getting maximum performance from electronic test gear is a never-ending quest for all successful technicians. To do this means selecting the test instrument that is best for the particular measurement and then performing each test in the most efficient manner. You'll find the factors to be considered for hundreds of tests and measurements in this book. For example, servicing solid state TV receivers can be a frustrating experience because special parts such as triacs, IC's, and "super modules" are encountered more and more. To make your servicing easier, there are scores of tests and measurements to help you service the most modern solid state TV receivers.

Here are just a few of the many tests and measurements this book will enable you to make ...

> How to quickly check an IF stage without instruments
>
> How to measure a CB rig's modulated output power with nothing but a dummy load and ammeter
>
> How to remove CRT face plate burns without extra equipment
>
> How to accurately test an alternator with an oscilloscope

Every chapter shows you how to get the *most* out of your tests and instruments. Some even show how to build circuits that will expand your test instrument's capabilities. Others show how to make reliable measurements with your VOM, VTVM, and scope that you may not have believed possible. You can be assured *every* important phase of practical electronics testing is covered.

In the past, we needed to remember only a few test measurement techniques to service almost every piece of electronics gear manufactured. But the transistor, integrated circuit, and many other solid state devices have changed all this. Today, we must test and measure an incredible variety of tube and solid state circuits, plus all of their associated components, creating the need for a modern encyclopedic guide to electronics testing and measurements such as this one.

As you can see, this book is a *practical* treasury of "real-life" testing and measurement techniques. It will quickly become one of the most valuable everyday aids you've ever used.

Robert C. Genn, Jr.

CONTENTS

5

8. Practical CB Radio Tests and Measurements (cont'd)

9. Electronics Tests and Measurements for Remote Control Equipment **181**

10. Practical Guide to Medium Power Transmitter Tests and Measurements **197**

11. Tests and Measurements for Antenna Systems and Transmission Lines **214**

Chapter 1

Time-Saving Tests and Measurements
For Semiconductors

No professional "high precision" equipment is needed to perform any of the tests in this chapter. Each test has instructions for selecting the right low-cost test gear and applying it to the circuit (or component) properly to obtain the information needed. Furthermore, every test uses a step-by-step approach that will help you improve your ability to use inexpensive test instruments during your daily work.

The entire chapter is aimed at simplifying semiconductor tests and measurements. Solid state devices have a large number of parameters that can be tested—particularly, transistors. However, only a few of these are useful during actual servicing. The only thing most of us are interested in is determining if a semiconductor is defective, and then looking through our spare parts box for a suitable replacement. Because this is all most of us really want to do, it's surprising how few pieces of test equipment are needed to check solid state components. Which procedure and test equipment you use will depend on what test gear you have on hand. To solve this problem, different ways of doing the same test are shown, with the simplest and least expensive way always given first.

1.1 SMALL SIGNAL DIODE TESTING—OHMMETER

Test Equipment:
Ohmmeter

Test Setup:
See Procedure

Comments:

Generally, it's best to use the R x 10 range to avoid exceeding the diode current rating. However, in some cases, you may have to use the R x 1 range in order to produce enough forward bias to cause full conduction. In most instances, when you switch your VOM to the ohmmeter function, the ground lead is positive and the red lead negative, but not in every case. Therefore, if you're not sure, it's best to check the lead polarities. Also, be careful with VTVM's because their polarity may be the exact opposite of VOM's.

Procedure:

Step 1. Connect the positive lead to the anode and the negative lead to the cathode. Your VOM should read a low resistance.

Step 2. Connect the negative lead to the anode and the positive to the cathode. You'll probably read infinity, or very near it, when checking silicon diodes. Germanium diodes should check out at several hundred k-ohms. A zero reading indicates the diode is shorted and should be discarded. The actual value of resistance measured means very little; the important thing is that a low resistance is measured in one direction and a high resistance in the other. This test is a simple way to identify which end is the cathode and which is the anode, in the event they are not marked. Assuming the diode checks out to be good, Step 1 identifies the anode and cathode. However, small signal diodes come in various cases or packages, a few of which are shown in Figure 1-1. You'll notice that some types have the cathode end marked with a stripe or band. Others use a diode symbol on the metal case, as shown in Figure 1-1 C. Diode identification numbers may begin with 1N (for example, 1N60, 1N35, 1N4001, etc.), or start with letters such as GE.

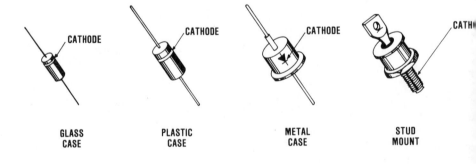

GLASS CASE   PLASTIC CASE   METAL CASE   STUD MOUNT

Figure 1-1: Diode packages

1.2   SILICON RECTIFIER PEAK INVERSE VOLTAGE TEST

Test Equipment:
High voltage power supply, microammeter, potentiometer (about 500 k-ohms, in most cases), and voltmeter

Test Setup:
Connect the variable resistor, microammeter, and silicon rectifier in series, as shown in Figure 1-2. The silicon rectifier under test must be connected in reverse polarity for the PIV test. Connect the voltmeter in parallel with the diode.

Comments:
The most important parameters of a silicon power supply rectifier are the maximum rated operating current and the peak-inverse-voltage (PIV). For example, you'll find that the specs for a General Electric GE-504-A TV silicon rectifier are 600 PIV and 1 amp forward operating current. The power supply used for the PIV test should be able to produce a voltage greater than the PIV rating of the diode under test.

Procedure:
Step 1. Connect the circuit as shown in Figure 1-2 and set the potentiometer to full resistance.

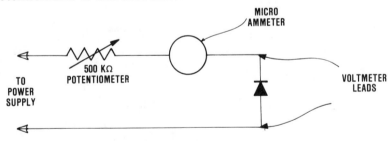

Figure 1-2: Circuit connections for a diode PIV test

Step 2. Turn on the power supply and adjust it to produce between 650 and 700 volts when testing a GE-504-A. Many other silicon rectifiers are rated for a PIV of 400 volts. In this case, set your power supply to about 500 volts.

Step 3. At this point, you may read a few microamperes, with only a slight increase as you slowly reduce the resistance. *Be careful!* As you continue to reduce the resistance, you'll see a sudden sharp increase in current flow. When the meter begins to show this rapid rise, *quickly* note the voltage reading on the voltmeter and then shut off the power supply. The voltage you read on the meter just before the sudden current increase is the PIV rating of the diode.

1.3   SILICON RECTIFIER FORWARD CURRENT TEST

Test Equipment:
Low voltage, high current power supply (for example, a Heathkit model IP-2715 or a B and K model 1640), 200-ohm resistor with a high wattage rating (for example, 1 ampere through 200-ohms produces 200 watts of heat), ammeter, and voltmeter

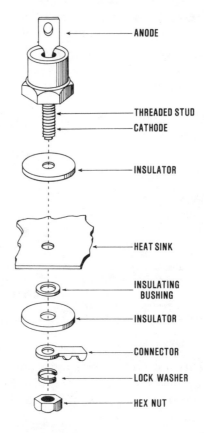

Figure 1-3: Stud mounted high current rectifier

Test Setup:

Connect the variable resistor, ammeter, and diode under test in series, as shown in Figure 1-4. Connect the voltmeter leads across the diode. Set the voltmeter to measure slightly over 1 volt. *Note:* The higher the diode current rating, the lower the voltage you'll measure across it; for example, about 1 volt for a 3 amp diode and 1.1 volt for a 1 amp diode.

Caution: Lower current ratings may have higher voltages across the diode (up to 90 volts or more in some cases).

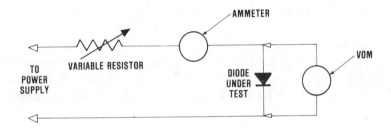

Figure 1-4: Test setup for checking the voltage drop, current, and power dissipated by a diode

Comments:

When testing silicon diode rectifiers, it's fairly easy when checking small axial-lead packages such as the "top hat" type GE-504 used for an example in test 1.2. However, you can get into much higher current drains. Two examples of the high forward currents that you may encounter in these diodes are: 1) Hep-153, having a 200 volt PIV rating and a current maximum of 15.0 amperes, and 2) RCA SK03500, having a 600 volt PIV and a current maximum of 12.0 amperes. Typically, they are stud mounted, as shown in Figure 1-3. The threaded stud usually is common to the case and electrically connected to the cathode. The anode is connected to the insulated terminal on the top.

Procedure:

Step 1. Slowly reduce the resistance value, watching both the ammeter and voltmeter.

Step 2. Generally, the current reading multiplied by the voltage reading shouldn't exceed slightly over one watt. However, in no case should this value exceed the wattage rating of the diode. If it does, the diode is assumed to be bad.

Step 3. Watch the voltmeter. It shouldn't read over 1 to 1.5 volts. Ordinarily, a silicon rectifier with a voltage reading over 1.5 volts should be discarded.

1.4 SILICON CONTROLLED RECTIFIER TEST

Test Equipment:
Ohmmeter and jumper lead

Test Setup:
See Procedure

Comments:
When checking an SCR using this procedure, you may find some that won't respond to Step 3 because of insufficient ohmmeter current. However, some of them are very sensitive so, if in doubt, always use a high resistance range.

Procedure:
Step 1. Connect your ohmmeter between the anode and cathode with the gate open. See Figure 1-5.

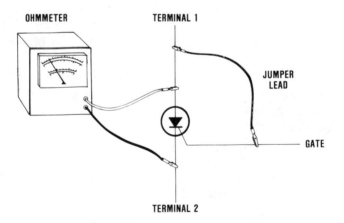

Figure 1-5: Ohmmeter connections for checking an SCR

Step 2. Measure the resistance between anode and cathode. It should read very high. Reverse the ohmmeter leads and, if it's operating properly, you'll read the same value of resistance (practically infinite). Reconnect the ohmmeter leads, as explained in Step 1, before making the next step.

Step 3. With the ohmmeter still in the circuit, place the short (jumper lead) between the anode and gate. You should see the ohmmeter reading drop to a low resistance.

Step 4. Remove the jumper wire, without any other changes. The ohmmeter should show the same resistance reading as you had in Step 3—a low resistance. *Note*: Once the diode begins conduction, the gate loses all control.

Step 5. Disconnect and then reconnect either of the ohmmeter leads. The resistance reading should jump back to its original high resistance reading. This completes the check.

1.5 PRACTICAL VARICAP DIODE TEST

Test Equipment:
Transistor tester or semiconductor analyzer

Test Setup:
Attach base and collector leads to the diode.

Comments:
Voltage variable capacitors, varicaps or varactors are PN junction diodes that perform like capacitors when biased in the reverse direction. Typically, the capacitance can be varied over a 10 to 1 range, with bias change from 0 to 100 volts and the current supplied by the bias supply not much more than a few microamperes, in most cases. One of the most important tests of a varicap is its reverse current leakage. This is an easy test to perform with a transistor tester.

Procedure:
Step 1. Connect the PN junction diode (varicap) to the transistor tester.

Step 2. Switch to both NPN and PNP positions. One will read forward and the other, reverse current.

Step 3. You should read no reverse leakage current on an ordinary transistor tester. Even the smallest reading indicates the varicap may be doing a poor tuning job and should be discarded.

1.6 ZENER DIODE TEST—IN CIRCUIT

Test Equipment:
Voltmeter

Test Setup:
None

Comments:
The voltmeter reading across a properly operating zener diode will remain practically constant no matter how much current is passed through the diode, provided it isn't driven beyond its operating range. There are two conditions that must be considered, to successfully check a zener. First of all, there must be at least a few milliamps flowing through the diode to keep a stable voltage. Second, an excessive curent flowing through it will burn it out. Just remember, the rated wattage of a zener is determined by the zener current at full conduction. Zeners come in a wide range of voltages .. from a few volts to a few hundred volts. They are available in power ratings of 250 mW, 500 mW, 1 W, 10 W, 50 W, and a few other ratings not included. Therefore, when making replacements, be sure to use an exact duplicate or, at least, use one that will stand up in the circuit you're working with.

Procedure:
Step 1. Measure the voltage across the load that is in parallel with the zener.

Step 2. If the voltage is quite a bit below the manufacturer's recommended value (or due to a comparison check, what you know to be normal), disconnect the zener. If your voltage reading jumps to well above the normal reading, it's an indication that the zener is leaking current and should be replaced. If there is no change, see Step 4. It's also possible that you'll read some voltage quite a bit above normal at all times. In this case, the zener is open.

Step 3. If you believe the zener is showing signs of an open circuit, disconnect it. If there is no change in the voltmeter reading, the zener is open.

Step 4. If your voltage reading is low when you measure across the load, it may mean an overload somewhere in the load circuit. Try disconnecting the zener and see if this changes the meter reading. If there is no change, you have an overload.

1.7   ZENER DIODE TEST—OUT OF CIRCUIT

Test Equipment:

Power supply, (this can be a radio, TV, or bench power supply just so long as it provides a voltage that is a little higher than the zener rated voltage) VOM and two resistors, a 5 or 10-ohm potentiometer and a 1 k-ohm fixed resistor

Test Setup:

Attach the zener cathode, which may be marked plus, to the positive lead of the circuit, as shown in Figure 1-6. Next, connect the zener anode, which may be marked negative, to the negative lead. Attach the 1 k-ohm current limiting resistor in series with the zener and the potentiometer.

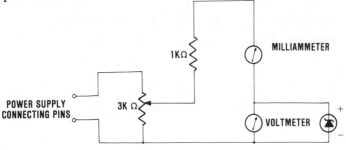

Figure 1-6: Test setup for checking the quality of a zener diode

Comments:

Read test 1-6 Comments before making this test.

Procedure:

Step 1. Turn on the power supply and measure the voltage across its output. The correct voltage will depend on the zener under test. For example, the zener will not start operating until you reach about 20% of its maximum load. For instance, suppose it's a 20 volt supply and you're going to test a 12 volt zener. Your first check is to see that you have 20 volts out of the supply (a few volts one way or the other is close enough).

Step 2. Turn the power off. Set the VOM to read milliamperes. Attach the circuit as shown in Figure 1-6. Set the pot to minimum, turn power on and advance the pot, all the while watching the meter carefully. When the reading starts to rise rapidly, you have reached the zener's *breakdown voltage* (also called *zener knee, avalaunch point,* and *zener voltage*). As you go up in voltage, the so-called *reverse current* will be extremely low until you reach the breakdown point, then the current will jump. If you increase voltage beyond this point, you should see only current increase. The voltage across the zener should remain constant. If it doesn't, the zener is bad.

1.8 PRACTICAL BIPOLAR TRANSISTOR TESTING

OHMMETER METHOD

Test Equipment:
Ohmmeter. *Caution*: Although it seldom happens, small signal transistors may be damaged when tested in the following manner with an ohmmeter on the R x 1 scale.

Test Setup:
For high resistance checks, set the ohmmeter to the R x 1000 range. Use the R x 10 range when checking forward resistance values.

Comments:
This measurement provides a quick way of checking whether a transistor is good or bad. Also, the measurement will tell you whether a transistor is NPN or PNP. The test is based on the fact that a transistor responds to an ohmmeter exactly as two diodes back-to-back. Typical packages used for bipolar transistors are shown in Figure 1-7.

Procedure:
Step 1. Connect the ohmmeter between the base and emitter terminals (Figure 1-8 A and B). If the transistor is good, it should read low resistance in one direction and high in the other.

Step 2. Connect the ohmmeter leads between the collector and base terminals and again it should read low resistance in one direction and high in the other.

Step 3. Using either polarity, connect the ohmmeter leads between the emitter and collector. You should measure a high resistance either way. A low resistance reading indicates the transistor has a leakage current and it probably should be replaced. These

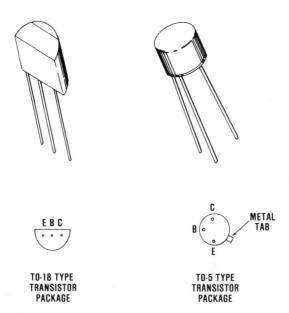

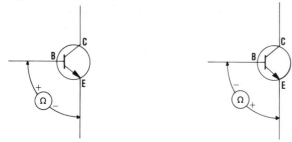

TO-18 TYPE	TO-5 TYPE
TRANSISTOR	TRANSISTOR
PACKAGE	PACKAGE

Figure 1-7: Typical small signal transistor packages

measurements will quickly identify a transistor as **PNP** or **NPN** (See Figure 1-8). It should be pointed out that *germanium power transistors* have *normal* leakage current of up to 100 or 150 microamperes, or even more. However, silicon power transistors should have practically no leakage current; therefore, very high resistance readings when checking reverse directions.

(A) LOW RESISTANCE INDICATES FORWARD BIAS   **(B) HIGH RESISTANCE INDICATES REVERSE BIAS**

Figure 1-8: This diagram shows the polarities and resistances you should find when checking a good NPN transistor. A PNP type transistor will be the exact opposite. It's important to know the actual polarity of the ohmmeter test leads in this test. If in doubt, check it with a voltmeter although some ohmmeters have red positive and black negative test lead terminals.

VOLTAGE METHOD—IN CIRCUIT

Test Equipment:
Voltmeter (high input impedance), screwdriver or jumper lead, and 10 k-ohm resistor

Test Setup:
See Procedure

Comments:
When making voltage measurements on transistor leads (common emitter or common base), you will find that a PNP transistor will have a positive voltage on its emitter, a negative voltage on its base, and a negative voltage on its collector. In other words, the base will be less positive than the emitter, and the collector should be the least positive of the three. However, in the basic NPN transistor, these polarities are reversed. The voltage between the base and emitter depends on the type of material used to manufacture the transistor. For example, you will find about 0.25 volts (anywhere from 0.1 to 0.4 volts) for germanium types (NPN or PNP), and about 0.6 (anywhere from 0.4 to 0.8 volts) for the silicon types. If your measurements are significantly different from these, the transistor is probably in need of replacement.

Procedure:
Step 1. Connect a voltmeter across the collector circuit resistor (common emitter configuration) as shown in Figure 1-9.

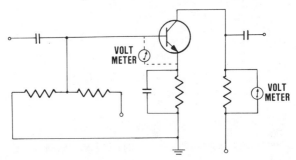

Figure 1-9: Voltage check points for a transistor amplifier

Step 2. Short the emitter to the base with a screwdriver or jumper lead. If the transistor is operating properly, the voltmeter reading should increase. If the voltmeter shows no change, there's a problem in the collector-emitter circuit.

Step 3. Measure the forward bias. If it's low, or you measure no bias, connect a 10 k-ohm current limiting resistor between the collector and base (common emitter configuration).

Step 4. Monitor the collector voltage and place a 10 k-ohm resistor in the circuit. This should decrease the collector voltage reading if the transistor is good. If the voltage does drop, start looking for a trouble in the bias circuit. See Figure 1-10.

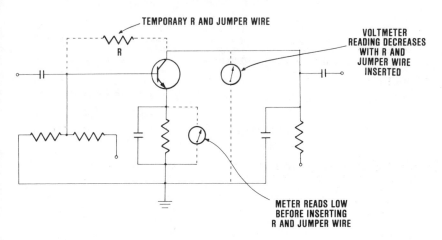

Figure 1-10: How to check a transistor by increasing bias

OSCILLOSCOPE METHOD—IN CIRCUIT

Test Equipment:

Oscilloscope, 6.3 VAC filament transformer, 270-ohm resistor, 3 jumper leads and 2 test leads

Test Setup:

Connect the transformer, resistor and test leads as shown in Figure 1-11

Procedure:

Step 1. Remove all power supply voltages from the transistor circuit under test.

Step 2. Connect the tester to the equipment as shown in Figure 1-11.

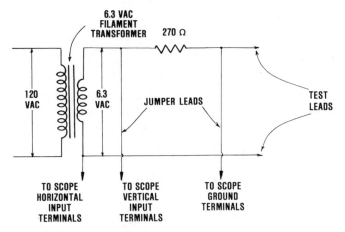

Figure 1-11: Transistor tester wiring diagram

Step 3. Connect the test leads to each of the transistor leads. Reapply power. You should see a sharp right angle on the scope as shown in Figure 1-12. The waveform may be inverted or exactly as shown. Either case indicates that the transistor is good.

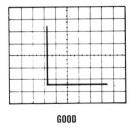

Figure 1-12: Waveform for a good transistor junction

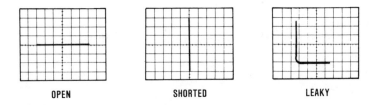

Figure 1-13: Waveforms for a bad transistor junction. Reversing the test leads should cause the waveform to flip.

Step 4. If the angle is rounded, the junction under test has leakage. A straight line means the junction is open or shorted. See Figure 1-13.

1.9   AF POWER TRANSISTOR COLLECTOR CURRENT MEASUREMENT

Test Equipment:
   DC milliammeter

Test Setup:
   See Procedure

Comments:
   Many audio frequency power transistors are mounted as shown in Figure 1-14. Notice that spring clips are mounted to the underside of the heat sink and bolts screw into the clips. It's important that the mounting bolts do not short to the heat sink during normal operation nor during the measurement to be explained. This type mounting usually uses one of the mounting bolts to complete the collector circuit, as shown in Figure 1-14.

Procedure:
   Step 1. Remove the bolt that has a terminal lug and wire between the clip and insulation. This is the lead that completes the

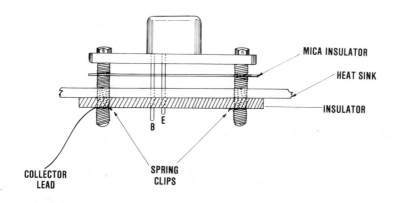

Figure 1-14: Power transistor mounting

collector circuit by permitting current to flow through the bolt to the transistor case, which is the collector in this type transistor.

Step 2. Connect the milliammeter in series with the collector lead wire and the other transistor mounting bolt shown in Figure 1-15.

Step 3. Apply power and read the collector current. It is important to note that your reading is the collector current of this single transistor and does not include any other component currents that may be in the circuit during normal operation.

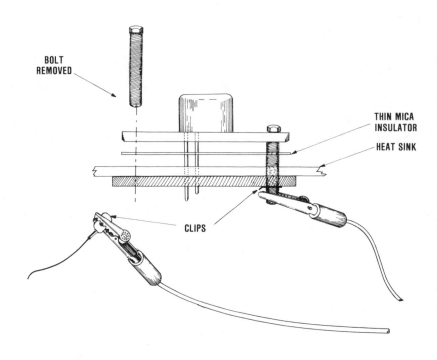

Figure 1-15: VOM connections for measuring a power transistor's collector current

1.10 JUNCTION FIELD EFFECT TRANSISTOR TEST

Test Equipment:
   Ohmmeter

Test Setup:
    See Procedure

Comments:
    Generally, when you purchase an **FET**, it comes with the leads twisted together or it's wrapped in metallic foil. The leads should be shorted at all times the unit is not in use to prevent a static charge from building up on the gate. Incidentally, ordinary kitchen aluminum foil wrapped around an **FET** works very well for storage. When you're working with these devices, always be sure everything is grounded because even the static charge of your body may damage it—especially in some of the older **MOSFET's**.

Procedure:
    Step 1. Connect the ohmmeter leads across the source and drain leads. You should measure a constant value from about 100 to 10 k-ohms. Reverse the ohmmeter test leads and you should measure the same value of resistance. If you don't, the **JFET** is bad.
    Step 2. Connect the ohmmeter leads between the gate and source. *Assuming an N-channel* **JFET** if the negative ohmmeter lead is connected to the source you should read a low resistance. Reverse the test leads and you should read a very high resistance—almost an open circuit. You will find the exact opposite when checking P-channel **JFET's**.
    Step 3. Connect the ohmmeter leads between drain and source. The gate-to-drain, or drain-to-source, resistance should be either low resistance or high resistance, depending on the ohmmeter lead polarity.

1.11   DUAL FET TEST

Test Equipment:
    Ohmmeter

Test Setup:
    See Procedure in Test 1.10

Comments:
    In dual type **FET's**, you'll find two source leads, two gate leads, and two drain leads. Test them in exactly the same way as was

explained in Section 1.10 (the preceding test), but check one at a time. If you find one section of the **FET** defective, probably the entire dual **FET** should be discarded although it's possible to use one section at a time, if you should want to save the half that is good.

Procedure:

See the Procedure given in Section 1.10. Follow the procedure for one-half of the dual **FET** and then repeat it for the other half.

1.12   METAL OXIDE SEMICONDUCTOR FET TEST

Test Equipment:

Ohmmeter, two 1 megohm resistors (½ watt), and jumper lead

Test Setup:

See Figure 1-16.

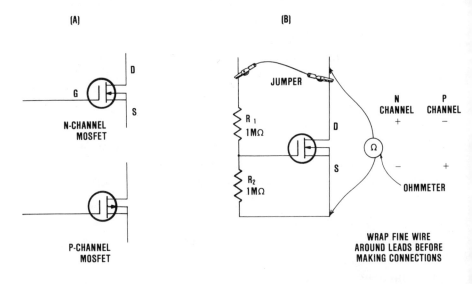

Figure 1-16: Test setup for checking an MOSFET (depletion mode shown)

Comments:

Unless the **MOSFET** is diode protected, its leads should be short circuited at all times except when installed in a circuit. Also, the transistor should never be removed from a circuit with power on since this may damage the unit. To disassemble a test setup, disconnect the ohmmeter and then remove the **FET**. It should be pointed out that although **MOSFET's** can be checked with an ohmmeter, it can be tricky because they are easily triggered by any stray transient voltage on the gate.

Procedure:

Step 1. Construct the circuit shown in Figure 1-16. Do all connections except the ohmmeter and shorting lead, then remove the shorting device, which may be a fine piece of wire wrapped around the **FET** pins.

Step 2. Connect the ohmmeter test leads across the source and drain. The negative lead is connected to the drain and the positive lead to the source, if the **FET** is a P-channel type. Use the reverse if it's an N-channel type—positive to drain, negative to source (See Section 1.15, Item 7). The schematic symbols are shown in Figure 1-16 A.

Step 3. Note your ohmmeter reading. It can be anywhere from 100-ohms to 10 k-ohms for a depletion **FET**. An *enhancement* type should measure infinity.

Step 4. Short between the top of R_1 and drain with the jumper lead. You should see the resistance reading drop to a lower value when checking either type **FET**.

Step 5. Connect and disconnect the shorting lead several times and you should see the ohmmeter reading increase and decrease as you make and break the connection. If the rise and fall are quite large, the transistor is probably good.

1.13   HOW TO TEST TRIACS

Test Equipment:

Ohmmeter and jumper lead

Test Setup:

In most cases, start off with your ohmmeter set to the R x 1 range. However, it's possible that the R x 1 range will produce too much current for a sensitive triac. To eliminate the problem, place 10 to 15-ohms of resistance in series with the ohmmeter test lead. On the other hand, you may encounter triacs that won't trigger with an ohmmeter. You'll have to have a larger voltage than the ohmmeter can produce to check these. But, they should respond in exactly the same way as described in the following procedure.

Comments:

Basically, the triac is a semiconductor type switch with three electrodes—two main terminals and a gate. For all practical purposes, it is equivalent to two **SCR's** connected in parallel (with one inverted), with a common gate. It provides switching action for either polarity of applied voltage and can be controlled in each polarity from the single gate electrode.

Procedure:

Step 1. Connect the positive lead of your ohmmeter to one of the main terminals (T_1, T_2) and negative lead to the other, as shown in Figure 1-17. The ohmmeter should read a very high resistance.

Step 2. Momentarily short between either terminal 1 or 2 and gate with a shorting lead. The ohmmeter reading should immediately drop to a much lower reading. Remove the shorting lead and there shouldn't be any noticeable change in resistance. This completes the test on one-half the triac.

Step 3. Connect the shorting lead between the gate and the other terminal. You should again read a very high resistance.

Step 4. Short the same two elements (gate to terminal) you used in Step 3 and the resistance should drop to a low reading. Remove the shorting lead and the ohmmeter reading should remain the same. This completes the test on the other half of the triac.

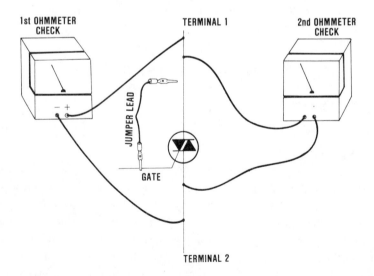

Figure 1-17: Triac test connections

1.14   HOW TO TEST A TUNNEL DIODE

Test Equipment:
   VOM, variable high voltage DC power supply (2 or 3 hundred volts), and 20,000-ohm resistor

Test Setup:
   See Figure 1-18.

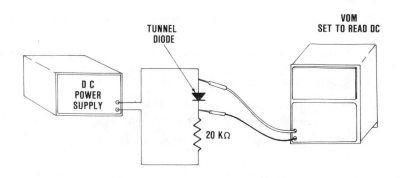

Figure 1-18: Test setup for checking a tunnel diode

Procedure:
   Step 1. Set the VOM to its lowest DC range. Typically, you'll be measuring about 0.5 volts.

   Step 2. Adjust the DC power supply so it produces zero volts on its output and connect the circuit as shown in Figure 1-18.

   Step 3. Adjust the power supply to produce an output voltage and watch the voltmeter reading. You'll see a sudden rise in voltage (from near zero to about 0.5 volts) when the diode switches, if it's performing correctly. *Note:* You may need several hundred volts out of the power supply to make the diode switch from a low to high voltage reading.

   Step 4. Next, reduce the power supply output voltage until you again see the diode switch. You'll see the voltmeter that is connected across the diode drop back down to the minimum reading. If the diode switches to a higher voltage (about 0.5 volts) in Step 3 and drops to some minimum voltage value in Step 4, it's an indication that the tunnel diode is good.

1.15 PRACTICAL TESTING PRECAUTIONS
FOR SOLID STATE DEVICES

1. Take care when using a standard VOM because it can permanently damage the small electrolitics used in much of today's equipment.

2. Dropping solid state components such as **FET's** can damage them.

3. As a general rule, do not exceed maximum current and voltage levels *even temporarily*.

4. In most cases, it's best to install the solid state component last in the circuit. Complete all wiring and attach the DC supplies with switch off, before making any test.

5. When disassembling a test setup, switch off all voltages before removing the solid state component under test.

6. When soldering, use long nose pliers to hold the component leads. Hold the lead until the solder is completely cool.

7. Many devices, such as the **FET**, are greatly affected by body capacitance that can cause errors in your testing. Always use a shielded probe and keep your fingers as far away as possible from the probe tip. *Don't* use an ordinary ohmmeter test lead held in your fingers, especially when working with operating **FET's**.

8. Ground your soldering iron tips and use a pencil type soldering iron.

Chapter 2

Tests and Measurements Needed to Service Radio Receivers

This chapter includes both solid state and tube-type radio receiver tests and measurements, and concentrates on the easiest and most practical way to make them with a minimum of low-cost equipment. However, professional work hasn't been forgotten. In many sections, you'll find several methods that are given in progressively greater detail, with more sophisticated test gear, that will insure optimum results for even the most demanding situation.

Whether you are an experienced technician interested in testing radio receivers more efficiently, or you're a beginner, this chapter will provide the information you need for specific types of tests and the specific test equipment you'll need to make each particular test using time-saving techniques. In every case, you may go directly to the test or measurement that describes the test procedure for a particular circuit or complete unit (Use the Table of Contents or Index to locate the desired test). The tests and measurements are complete by themselves, with only an occasional reference to other tests (by test number).

2.1 CHECKING THE LOCAL OSCILLATOR— TRANSISTOR RECEIVER

Test Equipment
VTVM or equivalent type voltmeter

Test Setup:
See Figure 2-1.

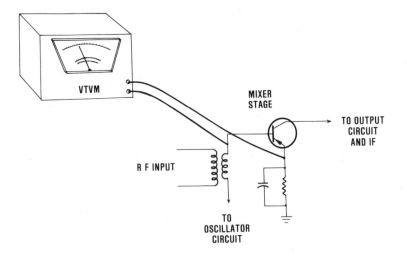

Figure 2-1: Test equipment setup for checking a transistor receiver local oscillator with a voltmeter

Comments:

An applied test signal is not required when checking a receiver's local oscillator because it is a self-regenerating circuit.

Procedure:

Step 1. Attach a high input impedance voltmeter between the emitter and base (See Figure 2-1).

Step 2. It is not important what value voltage you read. All you want is a reading. (Of course, the reading will be quite low—a few tenths of a volt). If the voltage remains fixed, the oscillator is not operating. If it changes a few tenths of a volt (increase and decrease as you tune), the oscillator is running.

2.2 CHECKING A LOCAL OSCILLATOR—
TUBE RECEIVER

Test Equipment:

VTVM or equivalent type voltmeter

Test Setup:

See Figure 2-2.

Comments:

Many times you'll find that an AM radio uses a pentagrid tube and local oscillator to make up the mixer. Three formulas that are very handy when servicing radio receivers are: LO = RF + IF; RF = LO –IF; and IF = LO –RF; where LO is the local oscillator frequency, RF is the radio frequency being fed to grid number one of the pentagrid converter, and IF is the intermediate frequency of the receiver. In almost every case, you'll find the LO and RF frequencies are beat together and the difference is 455 kHz (the IF).

Procedure:

Step 1. Connect the voltmeter hot lead to chassis ground (remember, we're measuring tube bias which, normally, is negative in respect to ground). Set the range switch to 10 VDC.

Step 2. Connect the voltmeter ground lead to the oscillator grid number one (See Figure 2-2). *Note:* If your voltmeter has a negative position lead switch, merely set the switch to the negative position and make the connections.

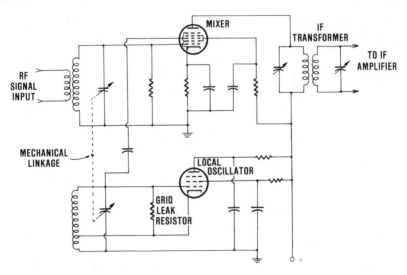

Figure 2-2: Pentagrid mixer with separate local oscillator. Many times you'll find the pentagrid converter acting as both local oscillator and mixer. However, it will have a slightly different wiring diagram.

Step 3. When the oscillator is running, there is a negative voltage developed across the grid resistor because of the grid leak bias design. For the same reason, when the oscillator is not operating there is no voltage developed across the grid resistor. Therefore, if you measure zero volts on the voltmeter, the oscillator is "dead." On the other hand, if you measure about 5 volts DC, plus or minus 3 volts, the oscillator is working.

2.3 IF AMPLIFIER OSCILLATION TESTS

VOLTMETER METHOD

Test Equipment:
   DC voltmeter

Test Setup:
   Connect a DC voltmeter across the load resistor of the receiver's second detector.

Comments:
   IF amplifier oscillation is caused by regeneration (regenerative feedback) which, in turn, may be caused by an open neutralizing capacitor or improper type transistor being used as a replacement. The symptom most frequently noticed is a distorted receiver audio output signal with poor intelligibility, particularly when tuned to a weak station. In severe cases, the receiver may not have any output signal or you may hear a high squeal.

Procedure:
   Step 1. Connect your voltmeter, set to read DC, across the second detector load resistor.

   Step 2. If the receiver is nonoperative and you read a high DC voltage across the resistor, it's a sure sign there is an oscillating IF amplifier. When the receiver is operating but has a distorted audio output, refer to the service manual or make a comparative check with a properly operating receiver of the same type to determine the correct voltage. If the IF amplifier is oscillating, you'll find a higher DC voltage present at the detector output.

   Step 3. To make a positive check, measure the bandwidth of the suspected IF amplifier (See Bandwidth Measurement). The sign of regeneration with this test is a comparatively narrow response curve. If the receiver oscillates with the AVC clamp voltage removed but stops with the clamp voltage connected, you probably have an open bypass capacitor on the AVC line.

SIGNAL GENERATOR METHOD—TRANSISTOR RADIO

Test Equipment:
RF signal generator, jumper lead and trimmer capacitor (10 - 350 pF)

Test Setup:
Connect the signal generator to the receiver antenna by using a few turns of wire. Set the generator to a frequency of about 1 MHz.

Procedure:
Step 1. Turn on the signal generator and receiver after you have the setup completed. Adjust the receiver volume control for a convenient level. Assuming an IF amplifier is oscillating, you may hear a squeal from the receiver at this point. However, there are other symptoms (See the preceding test).

Step 2. Take the jumper lead and short out the IF transistor you suspect, between emitter and base. When you do this, it cuts the transistor off by "killing" the forward bias.

Step 3. Connect the variable capacitor from the bottom of the IF amplifier output transformer to the base. See Figure 2-3.

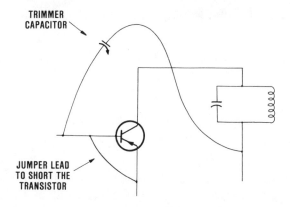

Figure 2-3: How to connect a trimmer capacitor during an IF amplifier oscillation test.

Step 4. Adjust the trimmer capacitor until you have minimum sound. If you can reduce sound to very little ... or no sound ... it's an indication that the IF amplifier was oscillating. Either it has no neutralizing capacitor or the neutralizing capacitor is open. In either case, place a fixed capacitor of the same value as the trimmer capacitor in the circuit. Or, possibly, you'll want to leave the trimmer capacitor in the circuit as is.

2.4 AM RADIO ANTENNA TEST—FERRITE CORE

Test Equipment:
   None

Test Setup:
   None

Comments:
   This test is to find the best position for the antenna windings on the ferrite core of an AM radio receiver. Figure 2-4 is an illustration of an AM radio ferrite core antenna.

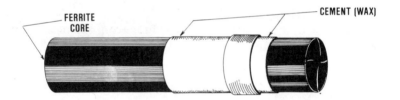

Figure 2-4: Antenna windings on an AM radio receiver ferrite core antenna

Procedure:
   Step 1. Free the antenna windings that you'll find wrapped around the ferrite core, with cement solvent, or loosen the wax. You can do this with your finger.

   Step 2. Tune the receiver to the low end of the dial (any weak station near 600 kHz) and slide the antenna windings back and forth until you have the best possible signal.

   Step 3. Tune the receiver to the high end of the dial (any weak station near 1500 kHz) and adjust the antenna trimmer capacitor for maximum volume.

Step 4. Repeat Steps 2 and 3 until the receiver is operating at peak volume at both ends of the receiver tuning dial. Then re-cement the antenna windings to the ferrite core. It may appear that there has been no change in the antenna coil position. However, even the smallest movement can make a considerable improvement in reception, especially in weak signal areas.

2.5   ANTENNA TEST USING A DIP METER

Test Equipment:
    Dip meter capable of tuning the frequency range of the receiver

Test Setup:
    Magnetically couple the dip meter to the receiver antenna.

Procedure:
    Step 1. Tune the receiver to the low end of its dial.

    Step 2. Tune the dip meter to the same frequency and look for the meter dip. If you get a good strong dip, it means the antenna is operating properly at this frequency.

    Step 3. Tune the dip meter to several other frequencies in the band the receiver is set to. Adjust the dip meter to the same frequency each time and look for the greatest dip (It's possible to tune up to a harmonic, so be sure to *look for the strongest dip*). If you get good dips at each test frequency, the antenna is good. If not, check the antenna connections, lead-in, and other components in the antenna system.

2.6   MEASURING RECEIVER SENSITIVITY

POWER METHOD

Test Equipment:
    Signal generator and voltmeter

Test Setup:
    See Figure 2-5.

Comments:
    Basically, what you're doing when you make a receiver sensitivity measurement based on power output is making an overall

receiver gain test without using internal noise as a reference. This measurement simply determines what level modulated signal is required at the receiver input to produce a certain power output (usually one-half watt, in better communications receivers).

Procedure:

Step 1. Connect the equipment as shown in Figure 2-5 and set the receiver RF and AF gain to maximum.

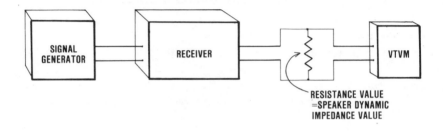

Figure 2-5: Test setup for radio receiver sensitivity measurement

Step 2. Adjust the signal generator to produce 1.4, 2, etc., volts across the load resistor. For all practical purposes, this is equal to 0.5 watts, assuming a load resistor of 4 or 8-ohms (using the formula $P = E^2/R$).

Step 3. Measure the RF voltage into the receiver. This value is the sensitivity of the receiver. Notice that no signal-to-noise ratio (S/N) is included. However, this check is very good to keep a running account of a receiver's performance or to compare one receiver to another.

PRECISION SIGNAL GENERATOR METHOD

Test Equipment:

High input impedance voltmeter, RF signal generator and dummy antenna, if needed

Test Setup:

See Figure 2-5.

Comments:

Sensitivity is expressed as the signal level required to produce a specified output signal having a specified signal-to-noise ratio.

Typically, it is the signal input needed to produce an output 10 dB above the receiver's internal noise.

To make a meaningful sensitivity measurement on a high gain communications receiver, the signal generator must have a good metering device. This is particularly important when working with high frequencies. Also, the voltmeter used during the measurement must be sensitive enough to read the receiver's internal noise when it is connected across the load resistor.

Procedure:

Step 1. Tune the receiver to a position on the dial where no station is detected. You should hear only noise with the gain control set at maximum.

Step 2. Disconnect the speaker from the receiver output and connect a resistor of the same impedance value as the speaker, in place of the speaker.

Step 3. Connect a voltmeter across the load resistor. If the receiver uses a 4-ohm speaker, you may have trouble reading a voltage across the resistor. In this case, and assuming there is an output transformer, move the voltmeter connections to the output transformer input winding, i.e., the transformer primary winding.

Step 4. A multiple band receiver should be set to its lowest frequency bandwidth and the RF sensitivity set to maximum.

Step 5. Adjust the audio volume control and voltmeter range switch until the VTVM reads 0 dB on the meter scale. This is your reference point with no signal input from the signal generator.

Step 6. Using a shielded cable, connect the signal generator to the receiver input. If the receiver usually is connected to a 50-ohm antenna, you can attach the signal generator output test lead directly to the receiver antenna input, because most signal generators have a 50-ohm output impedance. However, if the receiver has some other input impedance, you'll have to use a suitable dummy antenna. Several different types are shown in Figure 2-6. To select the correct type to use, you should consult the receiver service manual.

Dummy antennas are easy to make, but keep your leads as short as possible and mount the network in a metal box. An RF dummy load must be well shielded to prevent stray RF field pickup that can cause erroneous readings.

Step 7. Typically, signal generators are manufactured to

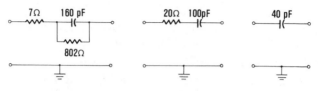

DUMMY ANTENNA FOR AUTOMOBILE RECEIVER TESTS
(CONSULT THE SERVICE MANUAL)

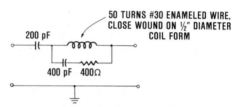

DUMMY ANTENNA FOR TESTING RECEIVER SETUP THAT WORKS INTO
A RANDOM LENGTH ANTENNA

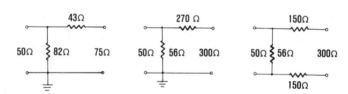

DUMMY ATENNA FOR MATCHING IMPEDANCES SHOWN
(CONSULT THE RECEIVER SERVICE MANUAL)

Figure 2-6: Receiver testing antenna networks

produce 30% modulation. However, if you are using external modulation or your signal generator has a variable internal modulation control, set it for 30% modulation at the frequency of the receiver.

Step 8. Next, adjust the signal output level of the signal generator to 1/10 of a volt (0.1V) with the signal generator step attenuator set at minimum attenuation. If your signal generator doesn't have a metered output, use a high input impedance voltmeter to make the setting.

After you've made the 0.1V setting, do not change the signal generator's variable attenuation control knob. At this point, all changes of output level must be done by changing the signal generator step attenuator. You should now have an output level of 0.1 V with the signal generator step attenuator set at zero dB, and the variable attenuator control set to maximum output.

Step 9. The step attenuator of a precision RF generator is frequently calibrated in dB. In this case, you have to convert the dB settings to microvolts. With steps divided by 10 and a reference of 0 dB = 0.1 V, we have the result shown in Table 2-1.

0 dB	=	0.1 VOLT
20 dB	=	0.01 VOLT
40 dB	=	0.001 VOLT
60 dB	=	0.0001 VOLT
80 dB	=	0.00001 VOLT
100 dB	=	0.000001 VOLT
		= 1 MICROVOLT = 1μ VOLT

Table 2-1:   Conversion of dB's to volts output when signal generator output attenuator is calibrated in dB's

Step 10. With the step attenuator set so the output voltage is 0.001 or 0.00001 volts (60 or 80 dB), adjust the signal generator frequency for a maximum reading on your voltmeter. If your voltmeter reading is off scale, increase the attenuation.

Step 11. The gain is read directly off the step attenuator, if the measurements fall on even numbers. The total attenuation required to reduce the output to your reference is the gain in dB. If the attenuator reading doesn't fall on even numbers, adjust the variable attenuator knob on the signal generator for a reading on your voltmeter that is as close as possible to 10 dB below the zero noise level. If the receiver output meter reads 13 dB, for instance, reduce the signal generator variable control to the 10 dB point. Notice this is 3 dB down. Therefore, using the formula, dB = 20 log voltage $_2$/voltage$_1$, we get, 3 = 20 log 0.707 or, approximately, 0.7. Multiplying this value by our reference (1 μV), we get 0.7 μV. This means the sensitivity of the receiver is 0.7 μV for 10 dB S/N.

Should the nearest setting be below 10 dB, you will have to turn the variable control on the generator up to where the receiver output meter reads 10 dB above zero. The voltage you read on the signal generator output, multiplied times your reference in μV, will give you the sensitivity of the receiver for 10 dB S/N.

2.7   AUDIO STAGE GAIN USING A VOLTMETER AND AF SIGNAL GENERATOR

Test Equipment:

AF signal generator, high input impedance voltmeter with demodulator probe, and a load resistor to replace the speaker

Test Setup:

Connect the AF generator set at 400 Hz (if you use 1,000 Hz, the following example will change values), across the second detector (AM demodulator) load resistor. Replace the speaker with a resistor 4, 8, or 16 ohms, capable of dissipating the receiver output load without changing value. Connect an AC voltmeter across the resistor you use to replace the speaker.

Procedure:

Step 1. Connect the equipment as was described under the heading Test Setup.

Step 2. Adjust the audio signal generator output to between one-half to one volt.

Step 3. Measure the voltage across the receiver terminating resistor.

Step 4. Let's say that the receiver is rated to deliver 2 watts to a load impedance of 8-ohms. A reading of 4 volts should be obtained with the one-half to one-volt input to the audio stages. To calculate the correct output voltage reading for various load resistors and powers, simply use the formula: voltage out = $\sqrt{(\text{resistance}) (\text{rated power})}$.

2.8   IF STAGE GAIN MEASUREMENT

VOLTMETER AND RF SIGNAL GENERATOR METHOD

Test Equipment:

RF signal generator, high input impedance voltmeter, demodulator probe and dummy antenna, if needed

Test Setup:

Set the voltmeter to measure a DC voltage and connect the demodulator probe across the AF demodulator (AM detector) load resistor. Set the range to a low scale. 0 - 1.5 volts, for example. Connect the signal generator to the input of the last IF stage.

Comments:

The gain of an IF stage depends upon its bandwidth. For instance, if the bandwidth of the stage is doubled, it will lose about one-half of its gain. Therefore, it would be impossible to achieve the gain given in the manufacturer's specs unless the bandwidth is set exactly as the service manual recommends. Reference to the service notes is the best method to find the correct gain for the receiver under test. Another good way is to make a comparative check in a normally operating receiver of the same type.

The overall gain of an IF amplifier strip is equal to the product of the individual identical stage gain. If three identical stages are used on the IF strip, the overall gain is the cube of the gain of a single stage. For example, ideally a triode transistor connected in a common emitter configuration will produce a gain of 270. Therefore, three identical stages will produce a voltage gain of approximately 270^3. With all this gain, the output of the last IF amplifier still will be only somewhere in the vicinity of a few tenths of a volt and, as you can see, it's all but impossible to measure the preceding stages with anything but a very good instrument.

Procedure:

Step 1. Connect the signal generator to the input of the last IF stage. If a DC blocking capacitor is needed, place a 0.01 μF capacitor in series with the signal generator's ungrounded test lead.

Step 2. Set the signal generator to the intermediate frequency of the receiver and adjust the variable output level control until you read 0.1 volts on the voltmeter.

Step 3. Move the signal generator test lead to the output of the last IF stage.

Step 4. Adjust the signal generator attenuator until you have a reading of 0.1 volt on the voltmeter. The difference in the readings is the voltage gain of the stage.

MEASURING IF STAGE GAIN IN dB'S METHOD

Test Equipment:

RF signal generator with its attenuator calibrated in dB's, high input impedance voltmeter and demodulator probe

Test Setup:

Connect the voltmeter demodulator probe across the receiver demodulator (AM detector) load resistor with the voltmeter set to measure DC volts. Set the range switch to a low range; 0 to 1 volt, for example. Connect the signal generator to the input of the last IF stage. If a DC blocking capacitor is needed, use a 0.01 μF capacitor in series with the generator's ungrounded test lead.

Procedure:

Use the same method employed in the preceding test, except when the signal generator attenuator is calibrated in dB's the gain of a stage is read directly off the attenuator range setting, if the measurements fall on even numbers. In cases that fall in between the even dB readings, use the generator's variable attenuation control. The total attenuation required to reduce the output to your reference is the gain of the stage in dB. However, since gain frequently is a voltage ratio, it may be necessary to convert. There is a voltage gain of 10 for each 20 dB (See Table 2-1). Incidentally, this measurement is basically the same as described in Section 2.6 (Precision Signal Generator Method).

IF STAGE GAIN MEASUREMENT—OSCILLOSCOPE METHOD

Test Equipment:

RF signal generator, oscilloscope, high frequency probe

Test Setup:

Connect the signal generator through a 0.01μF DC blocking capacitor or a small coil (a few loops of insulated wire will do), to the antenna of the receiver. Connect the scope probe to the input and output of the stage under test, as explained under Procedure.

Comments:

This test calls for a high frequency probe and it's important that you do not use a probe with excessively high capacitance. Many "home brew" and low-cost probes are simply small trimmer capacitors. In this case, adjust the probe capacitor to as small a value as possible to produce on the scope (as near as possible) a perfect reproduction of the signal generator output. Too much capacitance will detune an IF transformer and cause erroneous readings. Also, don't overload the amplifier (See Section 3.3, "Power Response Test" in Chapter 3).

It is important to stabilize the gain of a receiver before making the test. In most cases, you'll find the gain of an IF stage depends on the avc bias voltage developed at the second detector and fed back to the base of the IF amplifier transistor circuit. This is a variable voltage and should be set at a standard avc clamp voltage, for example; 1.5 volts. See the service manual for the value needed for the particular receiver you're testing.

Procedure:

Step 1. Let all equipment warm up for about one-half hour.

Step 2. Connect the RF signal generator test lead to the receiver antenna.

Step 3. Connect the scope high frequency probe to the input of the IF stage under test. You should see a waveform on the scope similar to the one shown in Figure 2-7.

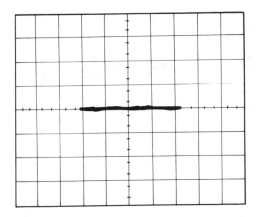

Figure 2-7: Oscilloscope display with high frequency probe connected to an IF amplifier input circuit

Step 4. Move the high frequency probe to the amplifier output circuit. The RF signal generator signal should appear larger and appear on the scope as shown in Figure 2-8.

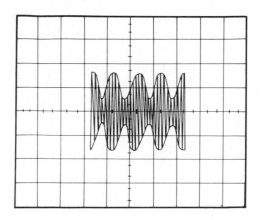

Figure 2-8: Oscilloscope display with high frequency probe connected to an IF amplifier output circuit

Step 5. Adjust the scope attenuator until you have the second waveform (See Figure 2-8) set to the same amplitude as the first one (See Figure 2-7). If the attenuator is calibrated in 10's and you make two steps in attenuation, the gain of the amplifier is 20. Remember, gain is a measured voltage ratio and 20 dB's is a voltage ratio of 10 (See Table 2-1).

2.9 MEASURING LOCAL OSCILLATOR FREQUENCY

Test Equipment:
Oscilloscope and calibrated signal generator

Test Setup:
See Procedure

Comments:
Sometimes when a receiver's local oscillator is running off frequency, it will result in a "dead" receiver. Or, if the local oscillator is drifting, it will cause you to have to keep retuning the receiver. It's very easy to check the local oscillator frequency and its stability with a signal generator and scope, as explained in the following procedure.

Procedure:

Step 1. Couple the scope probes to the local oscillator tuned circuit and adjust the scope until you see 1 to 4 cycles. Count the number of peaks in the waveform.

Step 2. Move the scope probe and connect it directly to the output terminals of the signal generator.

Step 3. Tune the signal generator until you see exactly the same number of cycles on the scope as you saw in Step 1. The reading you have on the signal generator dial is the frequency of the receiver's local oscillator.

Step 4. All you have to do to make a frequency drift test is monitor the scope pattern. The oscillator drift is determined by how much you have to adjust the signal generator to maintain the same number of cycles on the scope, and how often you have to make the adjustments. For example, if you have to correct five cycles per hours, that is the oscillator drift per hour.

2.10   AUTOMATIC VOLUME CONTROL TESTS

AVC VOLTAGE TEST — VOM METHOD

Test Equipment:
Voltmeter

Test Setup:
Connect the VOM between the avc line and chassis ground (or power supply negative terminal).

Comments:
Automatic volume control (avc) is a self-acting device that is supposed to maintain the output of a radio receiver or amplifier at a substantially constant level within relatively narrow limits, while the input voltage varies over a wide range. This same definition describes the operation of *automatic gain control* (agc). An avc system regulates the gain of the RF and IF stages and its output has to be a DC voltage. In a small transistor radio, it will be either negative or positive, depending on the type of transistor and which lead it is applied to: negative, for the **PNP** transistor, and positive, for an **NPN**, if the avc is applied to the emitter. When the avc is connected to the base of an

NPN transistor, it should be negative, and for a **PNP**, it's just the opposite, positive.

It's very easy to check whether the avc is supposed to be positive or negative. Simply look at the second detector diode polarity. If the cathode is connected to the avc line, it's positive. If the anode is connected to the avc line, it's negative.

A common avc trouble symptom is distortion on strong stations. There are several ways to check out the avc system with nothing but a voltmeter. In general, the higher the ohms-per-volt meter you use, the more accurate your measurements will be. Also, the most accurate reading will be obtained on the highest usable range of the voltmeter. However, when servicing transistor radios, in some instances, you'll have to use the voltmeter's lowest DC voltage range. For example, an avc problem is indicated where you measure an avc voltage of about 0.6 volts, if you measure it across the RF amplifier emitter resistor. Tube-type receivers will have much higher avc voltages—somewhere in the 0 to 10 volt region.

Procedure:

Step 1. Set the VOM DC switch to the highest usable DC range.

Step 2. In communications receivers that have a manual avc switch, check to see that the switch is set to its avc position.

Step 3. You must refer to the manufacturer's service data to make an ideal test. However, the avc output voltage is supposed to be a DC voltage (if it isn't, you'll have distortion in the RF and IF signals). Monitor the VOM and tune the receiver to various stations. As the signal increases in strength, the gain of the controlled stages is.thereby reduced by the increasing value of bias. Therefore, you should see a change in the *DC* voltage readings as you slowly tune from station to station. Incidentally, if the avc filter capacitor should open, the avc bias voltage will follow the rise and fall of the audio envelope and this may be heard as distortion in the receiver output.

Step 4. Finally, if you measure different avc DC voltages on adjoining ranges of your VOM, it's an indication the VOM is loading the circuit under test. This can result in considerably different voltage readings than recommended by the manufacturer. If you encounter this problem, it's possible to observe two of the different avc voltage readings and use the formula:

$$E = \frac{E_1 \, E_2 \, (R_2 - R_1)}{E_1 \, E_2 - E_2 \, R_1}$$

where E is the true avc voltage, E_1 is the VOM reading on the first range setting, R_1 is the VOM input impedance for the first range setting, E_2 is the VOM reading on the second range setting, and R_2 is the VOM input impedance for the second range setting. Or, you can use the following procedure.

AVC VOLTAGE TEST—HALF-VOLTAGE METHOD

Test Equipment:
   VOM and 500 k-ohm variable resistor

Test Setup:
   See Figure 2-9.

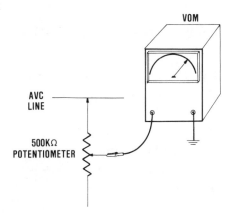

Figure 2-9: Measuring avc voltage using the half-voltage method

Procedure:

Step 1. Connect your VOM and a pot, as shown in Figure 2-9, and then set the VOM for about midscale, with the pot set at the zero resistance position.

Step 2. Adjust the pot resistance until the voltage is *exactly* one-half the reading you had in Step 1. Remove the pot and measure the resistance.

Step 3. Calculate the actual avc voltage by using the formula:

$$E_{avc} = \frac{E_1 R_p}{R_{meter}}$$

where E_{avc} is the actual avc voltage, E_1 is the first voltage measurement, R_{meter} is the ohm-per-volt rating of the VOM times the full-scale voltage, and R_p is the resistance value you measured across the pot that was required to reduce the voltage to one-half its previous value. For example, let's say that you're using a Heathkit model IM-105 VOM that has an input resistance of 20,000 ohms-per-volt, and is set to the 10 volt DC range. Next, assume that you measure 4 volts on the initial voltage reading, and a potentiometer resistance of 300,000 ohms when the VOM voltage reading has been reduced to 2 volts. The true avc voltage is:

$$E_{avc} = (E_1 R_p)/R_{meter} = (4)(300,000)/200,000 = 1,200,000/200,000 = 6V$$

2.11   AM RECEIVER TRACKING TEST

Test Equipment:
RF signal generator, AC voltmeter and dummy antenna (use the type specified by the manufacturer's service notes). See Figure 2-6 for typical dummy antennas used in radio receiver servicing.

Test Setup:
Connect an AC voltmeter across the speaker terminals. Connect an AM signal generator to the receiver antenna through a dummy antenna, as required by the manufacturer. Set the receiver volume control to maximum.

Comments:
Tracking is the maintainence of proper frequency relationships in circuits designed to be simultaneously varied by ganged operation. For example, the mixer and local oscillator circuits of a superheterodyne radio receiver are said to *track* if they maintain a constant frequency difference (usually the intermediate frequency equals the local oscillator frequency minus the input signal from the RF circuit) throughout the receiver tuning range. While it is impossible

to receive all stations with equal sensitivity (commonly, stations are weaker at the top end of the tuning dial, or the middle of it), the receiver will be operating at peak performance once you have the tuning circuits tracking as close as possible.

Procedure:

Step 1. Use a suitable dummy antenna and connect the signal generator to the receiver antenna. Set it to the high end of the dial—about 1400 to 1615 kHz. Some technicians use a standard fluorescent light. The light is an all-frequency signal source that will provide static anywhere you tune the dial. Simply make your tracking adjustments for maximum static in each of the listed steps. Weak radio stations also will do quite well, as explained at the end of this section.

Step 2. Using a weak test signal (with a fluorescent light, increase and decrease the noise level by varying the distance between the light and receiver), adjust the oscillator, RF, and antenna trimmers for maximum output. It's best to record your readings if you expect to check the receiver periodically because you can refer to them at some future date to see if the system is deteriorating.

Usually, the oscillator trimmer is in parallel with the oscillator tank and the antenna trimmer is in parallel with the antenna tuned input circuit. In automobile radio receivers, the oscillator tuning slug generally is set to some certain depth (see the manufacturer's service notes). As an example, 1⅜ inches frequently is used as the distance from the back of the coil to the tuning slug with the receiver set to the high end of the dial.

Step 3. Turn the signal generator dial to 600 kHz and tune the receiver for maximum output.

Step 4. Adjust the signal generator to a weak signal level output and tune the antenna and RF slugs for maximum receiver output. Perform the adjustments a few times, for best results.

For a quick tracking check, set the receiver to some weak station at about 600 kHz, with maximum volume. Adjust the *oscillator* trimmer for maximum signal level. Next, set the receiver to a weak station somewhere in the high end of the band. Adjust the *antenna* trimmer for maximum signal. Do this a few times and it should do the trick. If it doesn't, you probably need a more sophisticated tracking test such as that described in Steps 1 through 4. See Figure 2-10 for typical locations of antenna, RF, and oscillator tuning slugs in a mobile receiver.

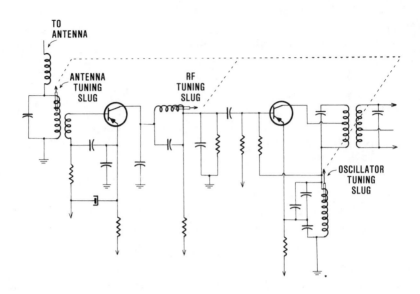

Figure 2-10: Typical location of antenna, RF, and local oscillator tuning slugs in a mobile receiver

2.12   FM DEMODULATOR TEST

Test Equipment:
   Sweep generator and oscilloscope

Test Setup:
   Connect your scope across the volume control and inject the sweep generator signal into the limiter if the receiver is using a discriminator, or into the last IF amplifier if it has a ratio detector.

Procedure:
   Step 1. Set your sweep generator to the receiver IF. Some receivers have an IF of 10.7 MHz and a bandpass of 200 kHz, but not all of them. In some cases, low-cost ones have a bandpass of only 50 kHz. Therefore, when in doubt, check the service notes.

   Step 2. Disable the local oscillator by placing a jumper wire across the oscillator coil.

   Step 3. Adjust your scope sweep frequency to 60 Hz. Adjust the sweep generator to sweep 200 kHz.

   Step 4. Adjust the demodulator by tuning the slug in the

transformer secondary until the curve is equally divided on both sides of the vertical line on the face of the scope.

Step 5. Adjust the primary slug until you have an S-shape curve at about a 45° angle, something like that shown in Figure 2-11.

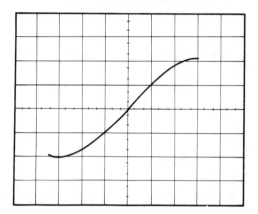

Figure 2-11:   An example of the curve you should see on your scope when checking an FM demodulator

Step 6. Set the horizontal sweep frequency of your scope to 120 Hz and you should see a new pattern on the scope. It should look like the one shown in Figure 2-12.

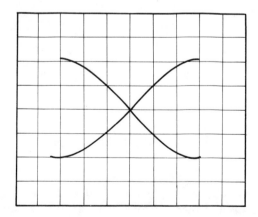

Figure 2-12: Scope pattern for last part of FM demodulator check

Step 7. Adjust the secondary coil slug until the two curves form, as nearly perfect as possible, an X pattern like the one shown in Figure

2-12. A perfect X pattern is an indication the demodulator is operating properly.

2.13   IF AMPLIFIER TEST FOR FM RADIO
WITH RATIO DETECTOR

Test Equipment:
Sweep generator and high input impedance VOM

Test Setup:
See Figure 2-13.

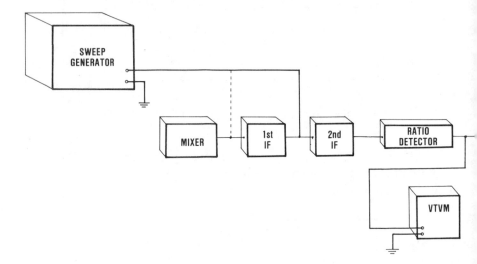

Figure 2-13: FM radio with ratio detector, IF amplifier test setup

Comments:
Basically, this test is an IF alignment check.

Procedure:
Step 1. Attach the sweep generator to the first IF amplifier input circuit. Short the receiver local oscillator by placing a jumper wire across the oscillator coil.

Step 2. Set the sweep generator to the receiver IF frequency (normally 10.7 MHz).

Step 3. Set the sweep generator to the correct sweep frequency.

This can be anywhere from 50 kHz up to more than 200 kHz, therefore you'll have to refer to the manufacturer's service notes. As a rule of thumb, very inexpensive receivers will have about 50 kHz bandpass and very expensive ones will be from 150 to 200 kHz, with some all the way up to 250 kHz, but this much is rare.

Step 4. Connect the voltmeter across one side of the ratio detector. The output voltage may be positive or negative. Either one is okay because it doesn't make any difference which one you can get.

Step 5. Simply check each IF amplifier (in inexpensive FM radios, you'll only find two) and adjust for the highest reading you can get. This will produce the best response. If you can't get a decent signal strength reading during this test, it's an indication the receiver is in need of a more sophisticated alignment.

2.14   STEREO AMPLIFIER POWER OUTPUT TEST

Test Equipment:
AC voltmeter or oscilloscope

Test Setup:
Connect the voltmeter or oscilloscope across the speaker coils.

Procedure:
Step 1. Set the gain controls of both channels to the same level.

Step 2. Measure the AC volts (rms) developed across each speaker voice coil with identical signal input. Remember, although the oscilloscope has the added advantage of letting you compare the two channels' waveforms, it reads a peak-to-peak voltage. Therefore, don't forget to convert the voltage measurement to rms voltage by dividing the peak-to-peak measurement by 2.8 before you do the next step.

Step 3. Compute the power out for each channel by using the formula, $P = E^2/R$, where R is the speaker impedance. Then make a comparison of the two. In most systems, they should be equal.

Step 4. To check the waveforms of each channel for distortion, use an identical input signal and monitor the output with your scope. A dual trace scope is ideal for this test.

Chapter 3

Audio Equipment Tests and Measurements

This chapter covers all the most important audio tests and measurements, and provides exact steps to follow in their application. Each test contains information that will help you select the right test gear, connect it to the circuit properly, and obtain the readings you need to determine if the equipment under test is operating within the manufacturer's specs.

To make things as easy as possible, more than one way to do a certain test is included. This makes it convenient to use whatever test gear you happen to have on hand. However, in general, the tests using less expensive test gear are less accurate—but easier and quicker to perform. On the other hand, there are times when more accuracy is important. In this case, you'll need high quality instruments and a slightly more complex way of doing the test. You'll find the simpler tests given at the beginning of each section, with the more sophisticated ones placed at the end of the chapter.

Specific instructions for making tests on particular audio equipment are purposely omitted because there is an amazing similarity in all audio units. For example, high-power musical instrument audio amplifiers use *precisely* the same output circuit as some low-power record player amplifiers. Therefore, you can use the following procedures on almost all audio equipment. About the only differences you'll find in testing methods are instrument range settings, due to different voltage and current levels.

3.1 FREQUENCY RESPONSE TESTS

SINE-WAVE GENERATOR AND AC VOLTMETER METHOD

Test Equipment:
AC voltmeter, sine-wave generator and terminating resistor

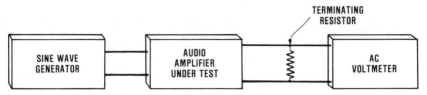

Figure 3-1: Test setup for amplifier frequency response measurement

Test Setup:

See Figure 3-1.

Comments:

Some signal generators don't have a metered output. In this case, you can use a second AC voltmenter to maintain a constant output level. However, both meters must have the same frequency response, to prevent erroneous readings. Also, to prohibit distortion and for·the safety of the amplifier, place a DC blocking capacitor between the signal generator and amplifier, using a value greater than the value for the capacitor in the amplifier input circuit (for example, the coupling capacitor connected to the base lead of a common emitter bipolar transistor amplifier).

Procedure:

Step 1. Start off with about 1/10 the highest normal operating power during warmup ... about 30 minutes to 1 hour ... and it's important you have a stable AC power line voltage (1% is considered excellent). Next, assuming the amplifier is operating with no overload, 1/3 maximum voltage output, low distortion, and has a fairly flat frequency response, make the frequency response measurement.

If you're not sure the amplifier has a comparatively flat frequency response, simply sweep the signal generator across the band and watch the AC voltmeter for a maximum reading. If there isn't much difference in readings, use 1 kHz for reference when you make the plot. The sketch in Figure 3-2 shows a typical frequency versus voltage output plot.

Should there be a substantial difference in readings when you sweep the amplifier bandpass, use the frequency with the highest voltage reading as your reference. The reason is to prevent overdriving the amplifier.

Step 2. Apply a constant amplitude 1 kHz signal voltage that will produce an output signal about 1/3 the maximum voltage output to the input of the amplifier.

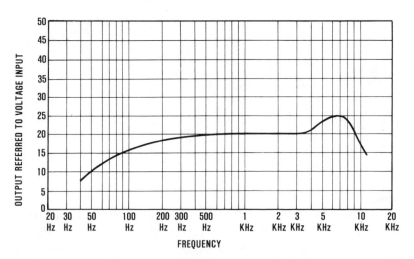

Figure 3-2: A typical frequency curve made using voltage and audio signal generator frequency settings with a reference of 1 kHz

Step 3. Measure the amplifier output voltage at several different frequencies (for instance, 1, 5, 10, 15 and 20 kHz) with the AC voltmeter.

Step 4. Make a plot of the output versus frequency and this is your frequency response curve.

SINE-WAVE GENERATOR AND OSCILLOSCOPE METHOD

Test Equipment:
Sine-wave generator, oscilloscope and terminating resistor, if needed, (refer to manufacturer's service notes)

Test Setup:
Connect the sine-wave oscillator to the input terminals of the amplifiler under test, and the oscilloscope to the output terminals, using a high impedance probe on the scope.

Comments:
Another approach to testing an audio amplifier frequency response is to use a sine-wave oscillator and ordinary oscilloscope equipped with a high impedance probe. It's possible to connect the scope directly; however, a high impedance probe is better in almost every case. The procedure is simple. To check the quality of an

amplifier, merely inject the sine-wave into the transistor input circuit and view its output on the scope.

Procedure:

Connect the scope test leads to the amplifier output and adjust its sweep frequency (time base) and other controls until you have two or three cycles appearing on the scope display. If the amplifier has a good frequency response, you'll see an excellent reproduction of the input sine wave at any frequency from about 50 Hz to 20 kHz ... assuming you are testing a hi-fi amplifier. Poor performance will be indicated by clipping of positive peaks, negative peaks, or both. Some clipping is sure to occur, but if the sine wave appears basically the same for all test frequencies, the amplifier is considered to have a flat frequency response. In Figure 3-3 you'll find some of the various scope presentations that indicate an amplifier is not responding properly. However, the oscilloscope must be free of distortion to reproduce a perfect copy of the test signal.

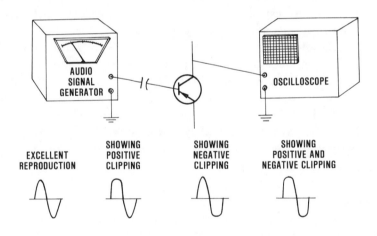

Figure 3-3: Distorted sine-wave patterns that may be seen on an oscilloscope, which indicate an audio amplifier is not responding properly during a frequency response check

SQUARE-WAVE GENERATOR AND OSCILLOSCOPE METHOD

Test Equipment:

Square-wave generator and oscilloscope utilizing a high impedance probe

Test Setup:

Connect the square-wave generator to the amplifier input terminals and the scope across the output terminals. See Figure 3-4.

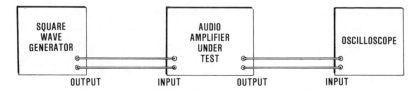

Figure 3-4: Block diagram of equipment setup for square-wave testing an audio amplifier

Comments:

In the foregoing measurment, we have used a sine-wave oscillator to excite the amplifier under test. However, a square-wave generator, oscilloscope, and high impedance probe are much easier to use when making frequency response checks than the point-to-point methods previously described. It should be pointed out that although the accuracy of this test is relatively poor, it is an excellent way to make a quick appraisal of an amplifier's performance.

To examine this matter in some detail, we must take into consideration the theory of square waves that states: "A square wave is made up of an infinite number of harmonics." This means that if an amplifier can't amplify *all* harmonics equally well, it will be impossible for the amplifier to reproduce a perfect square wave applied to its input. Therefore, any absence of harmonics due to poor amplifier frequency response is easy to see when observing the square wave displayed on an oscilloscope.

Procedure:

The output of the square-wave generator is applied to the input of the amplifier and the output of the amplifier is observed on the oscilloscope. Any deviations of the waveform seen on the scope from the input signal indicate a distinct absence of certain frequencies, as we have said. A 1,000 Hz square wave applied to the amplifier input will show how the amplifier performs from that frequency to the upper end of human hearing; a 50 Hz square-wave test signal will include the lower audio frequencies. If the amplifier can reproduce a square wave very near to a perfect replica of the square-wave generator output signal at both these frequencies, you can consider it to be free of frequency *and phase distortion* over the entire audio spectrum.

Deviations which may be seen on the scope that indicate specific problems in the frequency response of an amplifier are shown in Figure 3-5. At (A), the observed waveform indicates a loss in low frequency response. The problem may be an improper value (too small) coupling capacitor. Capacitors always decrease in value if they become partially open. At (B), the waveform shows a substantial drop in high frequency response. If you observe this pattern at certain high frequencies, the eighth harmonic of a 2 kHz test signal for instance, the amplifier can be considered to be linear up to 16 kHz (8 × 2 kHz = 16 kHz). The waveform at (C) indicates there is a higher gain at the lower frequencies. On the other hand, because the lower frequencies determine the shape of the horizontal portion of the square wave, it will sag in the middle if there isn't sufficient amplification at these frequencies, as shown in (A).

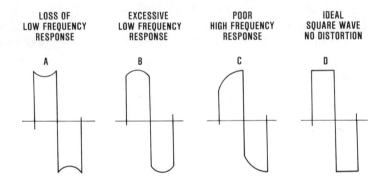

Figure 3-5: Scope presentations of amplifier frequency response deficiencies indicated by a change in a square-wave test signal

3.2 LINEARITY TESTS

SINE-WAVE GENERATOR AND OSCILLOSCOPE METHOD

Test Equipment:
Sine-wave signal generator and oscilloscope

Test Setup:
See Figure 3-6.

Comments:
To use a scope to make an audio amplifier linearity test, the first step is to check the scope linearity. Although this should be done in

almost all cases, it is a must in this test because a reference is required in order to make a comparison when evaluating the amplifier. To check your scope, connect the output test lead of the signal generator to both the vertical and external horizontal inputs of the scope, as shown in Figure 3-6.

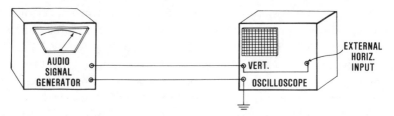

Figure 3-6: Test setup for checking a scope amplifier's linearity

The next step is to set the signal generator frequency to 1 kHz (audio tests used to be made using 400 Hz, but 1 kHz is used more frequently today). At this point, a diagonal line should appear on the scope screen. If the scope amplifiers . . . both vertical and horizontal . . . are linear, you'll see a perfectly straight diagonal line across the scope viewing screen, similar to the one shown in Figure 3-7. If the amplifiers

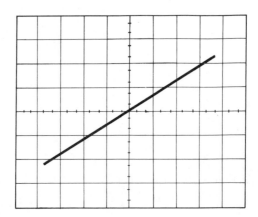

Figure 3-7: Correct oscilloscope reference linearity pattern

are not linear, you'll see a curved line. This is an indication that the scope needs some adjustments or repair. Refer to the manufacturer's instructions, in this case. After you have completed checking the scope amplifiers and are satisified they will not distort the test signal, proceed as follows.

Procedure:

Step 1. Connect the equipment as shown in Figure 3-8. The load resistor, R, must be capable of dissipating the amplifier output power without changing value, as explained in the section on Power Response Testing.

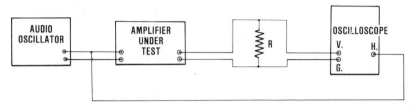

Figure 3-8: Test setup for checking an amplifier's linearity with an oscilloscope

Step 2. Adjust the signal generator output level until it drives the amplifier to its maximum undistorted power out. See the section on Power Response Testing.

Step 3. Finally, check the pattern on the scope. If you see exactly the same pattern that you saw on the reference pattern, the amplifier can be considered to be linear at the frequency being used to excite the amplifier (1 kHz, in this instance), all the way up to the maximum undistorted power. Notice this is not a frequency linearity test (See Amplifier Frequency Response Test).

SINE-WAVE GENERATOR AND STEP ATTENUATOR METHOD

Test Equipment:

Sine-wave generator, dB reading meter and step attenuator

Test Setup:

See Figure 3-9.

Comments:

The basic requirements are the same for this test as for all other tests; that is AC line voltage must be stable, all equipment should be left to warm up at about 1/10 the normal highest operating power, don't overdrive the equipment during measurements, and terminate the amplifier under test with the manufacturer's recommended load impedance, for example, a non-inductive load resistor. The test setup for this linearity measurement is shown in Figure 3-9.

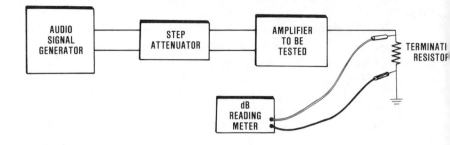

Figure 3-9: Equipment setup for measuring an amplifier's linearity using a step attenuator, audio signal generator and dB reading meter

To make a very accurate measurement of the linearity of an audio amplifier, the step attenuator should be capable of being varied in steps of 1/10 dB. This will permit you to determine the exact point when the amplifier departs from linearity. However, for general use, much larger steps may be tolerated.

Procedure:

Step 1. Set the step attenuator for an output signal level of about 20 dB below the amplifier's rated maximum power output (a power ratio of about 100 to 1).

Step 2. Adjust the step attenuator in 1 dB steps (or any other convenient value, if accuracy isn't a major concern), and watch the output signal level meter. For each step down in attenuation, you should see the same amount of increase on the output meter. For example, removing 1 dB of attenuation should produce a 1 dB increase on the output meter, if you're removing 1 dB of attenuation at a time.

Step 3. Finally, when the output meter reading starts changing less dB's than the dB changes made on the input signal attenuator, you have reached the amplifier's point of nonlinearity. As you can see, the finer the steps of attenuation, the more accurately you can read the exact departure from linearity.

3.3 POWER RESPONSE TEST—LOW AND MEDIUM POWER AMPLIFIERS

Test Equipment:

Sine-wave generator, scope, AC voltmeter and terminating resistor

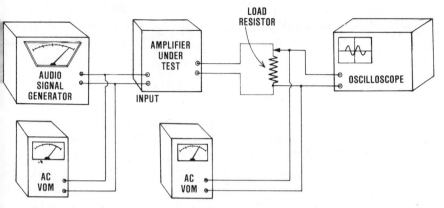

Figure 3-10: Test equipment setup for power response testing. *Note:* Although two meters are shown, it should be understood that only one meter is necessary if the signal generator has a metered output

Test Setup:

See Figure 3-10.

Comments:

Power response is defined as the frequency response capabilities of an amplifier running at, or near, its full rated power. There is a standard test procedure used by many service centers and manufacturers to measure an amplifier's power response. The measurement is made with the amplifier connected to a resistor with a very low reactive component (preferably no more than 10%). The resistor value is the same as the amplifier output impedance and must be capable of dissipating the full load of the amplifier while maintaining its resistance value very close to its rated value (ideally, within plus or minus 2%).

To do a professional job, all test equipment should be of very good quality. For example, the signal generator distortion should not exceed 20% of the measured distortion of the amplifier being tested and should be capable of maintaining the test frequencies within plus or minus 2%.

Procedure:

Step 1. Connect the signal generator to the amplifier input and set it to a 1 kHz test frequency.

Step 2. The scope and AC voltmeter are connected across the load resistor, as shown.

Step 3. Set the volume control to maximum and follow the manufacturer's specs for all other controls. Most of the time the other controls—for example, tone controls—are set to the center position, except for special circuits such as bass boost, which is usually set to off.

Step 4. Apply power and adjust the signal generator until you see a sine-wave pattern on your scope.

Step 5. Watch the waveform on the scope as you keep increasing the input signal level (probably, you'll have to reduce the scope vertical gain to keep the pattern on the screen). When you see the first signs of distortion on the sine wave (you'll see flattening peaks), reduce the signal generator output level until the clipping of the peaks just barely disappears. This is the amplifier's maximum undistorted power output level.

Step 6. Now we come to the reason for the two AC voltmeters, which is two-fold. One, it is necessary to hold the input voltage constant over the frequencies you decide to check (this isn't a problem with a calibrated attenuator, if the AF generator has one), and two, to compute the amplifier power out, you'll need to know the rms voltage developed across the load resistor. To do this, simply use the formula, $P = E^2/R$, where P is watts, E is the voltage measured across the load resistor and R is the value of the load resistor.

3.4   HIGH POWER AMPLIFIER TEST

Test Equipment:
Sine-wave generator, four 8-ohm load resistors, and a high input impedance AC voltmeter

Test Setup:
See Figure 3-11.

Comments:
Not long ago an audio amplifier was considered to be a high power amplifier if it could produce 50 watts of continuous power when driving an 8-ohm speaker. Today, there are amplifiers that generate 100 watts or more. Therefore, sometimes it's necessary to measure power output using a voltage divider terminating resistor network. A simple voltage divider load resistor circuit is shown in Figure 3-11.

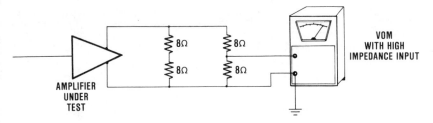

Figure 3-11: Voltage divider network that can be used to measure the output of a high power audio amplifier

Procedure:

Step 1. Let's assume the amplifier is rated at 100 watts of output power and has an 8-ohm output impedance. To make the power measurement, connect four 8-ohm resistors, as shown in Figure 3-11.

Step 2. Connect your VOM across one of the 8-ohm resistors, as shown. In this case, you should read approximately 1.767 volts rms on your AC voltmeter.

Step 3. Now, using the formula, $P = E^2/R$, we get 25 watts. However, the measurement was taken across the voltage divider so we'll have to multiply by 4 to get the true power reading. The end result is 100 watts, which indicates that the amplifier is performing according to specs.

3.5   INPUT SIGNAL LEVEL TEST

Test Equipment:
Sine-wave generator, scope, load resistor and AC voltmeter

Test Setup:
See Figure 3-10.

Comments:
This test uses exactly the same setup as was explained for Power Response Testing (See Figure 3-10), however, you don't need the voltmeter on the output circuit.

Procedure:
Step 1. Adjust your signal generator for maximum undistorted power, as explained in the section on Power Response Testing.

Step 2. Next, read the voltage across the amplifier input circuit, or take the reading from the signal generator's calibrated attenuator. If you can produce about 400 millivolts on the AC voltmeter without causing clipping of the sine-wave output signal you see on your scope, you can consider the amplifier to be in good working order. This test is called a *sensitivity test* by some manufacturers, and its purpose is to check the amount of input signal needed to drive the amplifier.

3.6 TESTING AN AUDIO SINE-WAVE GENERATOR FOR HARMONICS

Test Equipment:

AC voltmeter, capacitor, resistor and the audio sine-wave generator under test

Test Setup:

See Figure 3-12.

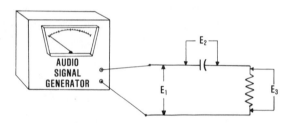

Figure 3-12: Equipment setup for testing an audio sine-wave generator for harmonics

Comments:

This test is based on the fact that the voltage reading (E_1) will be an exact 45° diagonal of the voltages E_2 and E_3, when plotted, if $X_C = R$ *and there are no harmonics.*

Procedure:

Step 1. Select the values of R and C that will produce approximately the same voltage readings at your chosen frequency.

Step 2. To compute the value of the capacitor, simply use the formula $C = 0.159 / (F_r X_C)$, where F_r is the frequency you chose to use, and X_C is the same value as the resistor.

Step 3. After you've made all three voltage readings, draw a plot like the one shown in Figure 3-13.

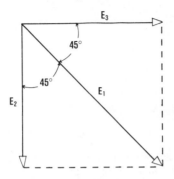

Figure 3-13: A plot similar to the one shown indicates the audio signal generator is free of harmonics

If you find E_2 and E_3 are equal in length and the E_1 length completes the diagonal line, it's an indication that the audio signal generator is free of harmonics. But if you can't plot a rectangle as shown, the signal source is producing harmonics.

3.7   AUDIO FREQUENCY MEASUREMENTS

SIGNAL GENERATOR AND AC VOLTMETER METHOD

Test Equipment:
Good quality AC voltmeter and sine-wave signal generator

Test Setup:
See Figure 3-14.

Comments:
There are numerous ways to measure an audio frequency. Without question, using a digital frequency counter is the easiest and most accurate of all methods. However, it's possible to make a fairly accurate audio frequency measurement using nothing but a good quality audio signal generator and AC voltmeter. The test setup for the measurement is shown in Figure 3-14.

Procedure:
Step 1. Adjust the output of the signal generator until you have

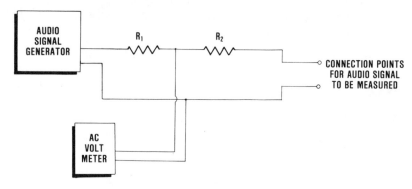

Figure 3-14: Circuit connections for measuring an audio frequency with an
AC voltmeter and audio signal generator (See procedure for
R_1 and R_2 values)

a center scale reading on the AC voltmeter without the unknown signal
being connected to the circuit.

Step 2. Next, connect the unknown frequency to the connection
points shown in the diagram.

Step 3. Tune the signal generator and watch the AC voltmeter.
As you tune the generator, you'll see the AC voltmeter needle start to
pulsate. The faster the meter swings back and forth, the farther off
frequency you are. The slower, the closer the signal generator is to the
unknown frequency.

Step 4. When the generator is set at exactly the same frequency
as the unknown frequency, you'll see no movement of the meter needle
(zero beat). Finally, read the frequency off the audio signal generator
dial. This reading is a quite accurate value of the unknown frequency.

The two resistors (R_1, R_2) shown in Figure 3-14 can be any two
resistors of equal value, that you happen to have on hand, that are
somewhere near 1,000 ohms (plus or minus a few-hundred ohms). The
greatest precision can be gained by setting the standard signal
generator output to about ten times the level of the unknown signal
amplitude, although lesser amplitudes can be used without too much
loss in accuracy. Also, it's good practice to check the standard signal
generator to 60 Hz and watch for slow AC meter pulsations. More than
likely, there won't be any. However, it's best to check, especially if you
can't bring the equipment to zero beat.

OSCILLOSCOPE METHOD

Test Equipment:
Oscilloscope with a calibrated time base

Test Setup:

Apply unknown signal to the vertical input of the scope.

Procedure:

Step 1. Adjust the oscilloscope sweep rate until you see one motionless cycle on the face of the scope.

Step 2. Note the width of the cycle by counting the number of divisions it fills on the scope graticule and find the time period by checking the settings of the oscilloscope sweep control.

Step 3. Now, simply use the formula, frequency = 1/time period. For example, let's say you count 4 graticule divisions for one cycle on the screen. Set the sweep control of the scope at 10 m sec per division and the scope multiplier at times 1. In this case, time = 4 x 10 = 40 m sec or 0.04 seconds; therefore, dividing this into 1, we get 25 Hz.

OSCILLOSCOPE AND CALIBRATED SINE-WAVE GENERATOR METHOD

Test Equipment:

Oscilloscope and calibrated standard sine-wave generator

Test Setup:

Apply the standard signal generator output to the external horizontal input of the scope. Connect the unknown signal to the vertical input of the scope.

Procedure:

Step 1. Adjust the amplitude of the standard signal generator until it is the same as the unknown signal. You'll know when the amplitudes are equal and the two frequencies are exactly the same (and 90° out of phase), when you see a perfect round circle on the face of the scope.

Step 2. Read the frequency off the standard signal generator and this is the frequency of the unknown signal. If the unknown frequency is an exact multiple of the standard frequency, 2 times, 3 times, or 4 times as high, you'll see the patterns shown in Figure 3-15 A, B, and C. But, should the unknown frequency be a submultiple, 1/2, 1/3 or 1/4 of the standard, you'll see the patterns shown in Figure 3-15 D, E, and F.

Figure 3-15 shows only a few of the many patterns that are possible to create using different frequency ratios. However, a

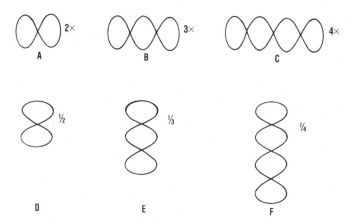

Figure 3-15: Scope patterns and frequency ratios when two sine-wave signals are out of phase. The phase shift is produced automatically as you adjust the frequency.

frequency ratio of 10 to 1 is about the highest ratio you can use because it becomes impossible to count the loops on the scope. For example, if you can count ten loops along the top of B, it indicates that the unknown frequency is ten times as high as the reading on the signal generator you're using for a reference.

DUAL TRACE OSCILLOSCOPE AND SIGNAL GENERATOR METHOD

Test Equipment:
    Dual trace oscilloscope and signal generator

Test Setup:
    Unknown signal to channel A. Reference signal generator to channel B.

Procedure:
    Step 1. Adjust the scope for a single steady sine-wave cycle on the channel B trace.

    Step 2. Next, adjust the wave pattern on channel A until you see 1, 2, 3, or more, cycles on the screen. If you see only one cycle on each channel, the two frequencies are the same. Two cycles on channel B means the ratio is 2 to 1, three cycles is 3 to 1, and so on. This method is just as accurate as your standard signal generator, which is also true of the preceding method.

3.8 HOW TO CONVERT POWER OUT
FROM VU's TO WATTS

It isn't the least bit unusual to find that manufacturer's specs give an amplifier output in volume units (VU's), but frequently it's necessary to convert VU's to watts. An example of how to do this can best be shown with a practical in-shop measurement.

Let's assume your VU meter reads 30 VU and you have exactly 0 VU (1 milliwatt in 600-ohms) input signal. Now, all you have to do is use the following formula and do your work as shown.

output power = antilog VU/10 = antilog 30/10 = antilog 3 = 1,000

Your answer will be in the same units as your reference level; milliwatts in this case. So the answer is 1,000 milliwatts, or simply 1 watt.

3.9 HOW TO CONVERT VOLTAGE READINGS TO dB's

Experience is really the best teacher of dB/volt conversion methods. But it won't be long, if you follow these tips, before you will be able to convert the readings, even with amplifiers that have unequal input and output impedances. Most electronics technicians are familiar with calculating voltage gain in dB's for an amplifier that has equal input and output impedances. However, to review quickly, it requires two AC voltage measurements; 1) the voltage on the amplifier input E_1; and 2) the voltage across the output load termination E_2. Then simply use the formula, $dB = 20 \log_{10} (E_2/E_1)$. Table 3-1 shows voltage and current ratios converted to dB's. The decibel calculations have been rounded off slightly, but they are sufficiently accurate for all the work most of us will encounter.

If the amplifier has unequal input and output impedances and you want to calculate the voltage gain in dB's, you have to use the formula:

$$dB = 20 \log_{10} (E_2 \sqrt{Z_1} / E_1 \sqrt{Z_2})$$

To find E_1 and E_2, measure the voltage at the amplifier input (E_1) and measure the voltage across the output load termination (E_2). Next, the value for Z_1 is the amplifier input impedance and Z_2 is the output load impedance.

Impedance depends on the frequency you use during the measurement, because both X_L and X_C vary with frequency.

VOLTAGE AND CURRENT RATIO	DECIBELS	VOLTAGE AND CURRENT RATIO	DECIBELS	VOLTAGE AND CURRENT RATIO	DECIBELS
1.0116	0.1	1.4962	3.5	12.589	22.0
1.0233	0.2	1.5849	4.0	15.849	24.0
1.0351	0.3	1.6788	4.5	19.953	26.0
1.0471	0.4	1.7783	5.0	25.119	28.0
1.0593	0.5	1.8836	5.5	31.623	30.0
1.0715	0.6	1.9953	6.0	39.811	32.0
1.0839	0.7	2.2387	7.0	50.119	34.0
1.0956	0.8	2.5119	8.0	63.096	36.0
1.1092	0.9	2.8184	9.0	79.433	38.0
1.1220	1.0	3.1623	10.0	100.000	40.0
1.1482	1.2	3.5481	11.0	125.89	42.0
1.1749	1.4	3.9811	12.0	158.49	44.0
1.2023	1.6	4.4668	13.0	199.53	46.0
1.2303	1.8	5.0119	14.0	251.19	48.0
1.2589	2.0	5.6234	15.0	316.23	50.0
1.2882	2.2	6.3096	16.0	398.11	52.0
1.3183	2.4	7.0795	17.0	501.19	54.0
1.3490	2.6	7.9433	18.0	630.96	56.0
1.3804	2.8	8.9125	19.0	794.33	58.0
1.4125	3.0	10.0000	20.0	1000.00	60.0

Table 3-1: Voltage/current ratios converted to decibels, and vice versa

Therefore, any impedance value you come up with is valid only for that frequency at which it was claculated or measured. Typically, 1,000 Hz is used as a reference frequency for measurements between 20 and 20,000 Hz.

3.10 HOW TO MEASURE POWER TRANSFORMER MAGNETIC COUPLING IN A TUBE-TYPE AMPLIFIER

Test Equipment:
     AC voltmeter (preferably with a low millivolt full-scale range setting) and a terminating load resistor

Test Setup:
     Connect a load resistor across the output of the amplifier (See manufacturer's service notes for recommended value) and then connect an AC voltmeter across the resistor.

Comments:

If you are having trouble with 60 cycle hum or second harmonics (120 cycle hum, in audio circuits), it's a good chance that you have magnetic coupling problems (assuming the power supply filter capacitor has been checked and found to be good). In some cases, the power transformer may be magnetically coupled to the chassis. To measure the amount of coupling in a situation like this is very easy, if you're working on a tube-type amplifier. Following, is a step-by-step procedure for making the measurement.

Procedure:

Step 1. Pull all tubes including every tube in the power supply.

Step 2. Remove the power transformer mounting screws.

Step 3. Place a terminating load resistor on the output of the amplifier.

Step 4. Connect your VTVM (or any other high impedance AC voltmeter) across the amplifier load resistor.

Step 5. Apply power to the input of the power supply; i.e., energize the system with the power line switch.

Step 6. Adjust your AC voltmeter for a convenient reading.

Step 7. Adjust the transformer to several different positions. Note the position that produces a minimum reading on your voltmeter. Once you find the spot that produces the minimum voltage reading, leave the transformer at that position.

Step 8. Turn all power off and replace all tubes. Again re-energize the system and take a measurement across the load resistor. Ideally, you should find the hum measurement to be very near zero, or, better yet, zero.

3.11 SINGLE-CONE SPEAKER
IMPEDANCE MEASUREMENTS

VOM AND AUDIO-SIGNAL-GENERATOR METHOD

Test Equipment:

High impedance VOM, audio-signal generator, and a 20-ohm variable resistor

Test Setup:

See Figure 3-16.

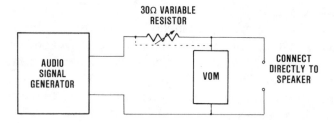

Figure 3-16: Circuit for measuring the impedance of a dynamic loud speaker

Comments:

The impedance of a dynamic speaker is usually specified for a certain frequency—typically 400 Hz. You'll find that the DC resistance measured with an ohmmeter is very close to the impedance of the speaker when it is excited with a 400 Hz signal. However, decrease the test frequency and you'll find that the dynamic impedance will increase sharply. On the other hand, increasing the frequency to above 400 Hz will cause the impedance to rise to a high value, but at a lower rate of climb. If you plot a curve of the impedance at different frequencies, it will look like the one shown in Figure 3-17.

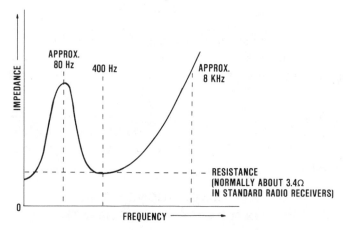

Figure 3-17: Typical single-cone speaker frequency versus impedance plot. It is very apparent that you must use a frequency fairly close to 400 Hz to measure the speaker impedance, or you'll end up with a value much too high.

Procedure:

Step 1. Measure the DC resistance of the speaker with the VOM.

Step 2. Connect a variable resistor of about twice the value of the DC resistance in series with the speaker. *Note*: Using twice the value will require a resistor of about 10-ohms, for most radio receivers. However, using a 20-ohm resistor will work with almost any speaker impedance. Also, in many of today's high power amplifiers, the watts rating of the resistor may be important if you use the amplifier output as your signal source. In this case, drive the amplifier with the signal generator set at 400 Hz. Other than this change in hookup, use the same setup as shown in Figure 3-16.

Step 3. Set the signal generator to 400 Hz. Connect the VOM to point A and, with the variable resistor, adjust for a convenient voltage reading.

Step 4. Move the VOM to point B and adjust the variable resistor until the voltage reading in Step 3 drops to one-half its previous value. When the voltage at point B exactly equals the voltage at point A, the impedance of the speaker is equal to the value of the resistor.

Step 5. Remove the variable resistor and measure its resistance with your ohmmeter. Your resistance reading is the impedance of the speaker. Incidentally, if you have a resistance box, it can be used in place of the variable resistor to save a few steps. But be sure it can handle the power, if the speaker is connected to an amplifier during the measurement.

OSCILLOSCOPE AND AUDIO-SIGNAL GENERATOR METHOD

Test Equipment:
Oscilloscope, audio-signal generator and a 20-ohm variable resistor

Test Setup:
See Figure 3-18.

Procedure:
Step 1. Connect the variable resistor in series with the speaker. Apply a 400 Hz signal to the speaker, through the resistor.

Step 2. Connect the scope vertical input lead to the signal generator's output terminal.

Step 3. Connect the scope horizontal lead to the speaker's ungrounded input terminal.

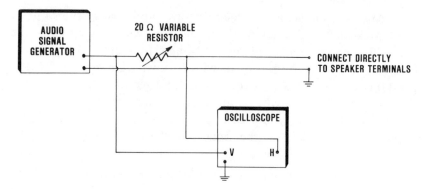

Figure 3-18: Connection for measuring a single-cone speaker impedance with an oscilloscope and audio-signal generator

Step 4. Set the sweep frequency of the scope to view a 400 Hz signal and adjust the vertical and horizontal level controls until you have a line about 3 or 4 inches long on the viewing screen. The line will probably slant one way or the other, but this is not important. Either way is okay.

Step 5. Adjust the variable resistor and scope controls until you have a line on the viewing screen that slants as close as possible to 45°.

Step 6. When you have the line slanting at 45°, the value of the resistor is equal to the impedance of the speaker. Simply measure the value of the resistor and this value is the speaker impedance at 400 Hz.

3.12   SPEAKERS IN PHASE TESTS

PHASING METHOD WITHOUT TEST EQUIPMENT

Test Equipment:
    None

Test Setup:
    See Procedure.

Comments:
    Most loudspeakers have coded terminals; sometimes a red dot or a plus sign, and sometimes a red fiber washer under the positive terminal. However, some are not marked. This simple test will work in cases of this type, as well as make a quick check of speaker phasing. A more thorough check is given in the next section.

Procedure:

Step 1. Connect the two speakers to the stereo amplifier output terminals.

Step 2. Reverse the speaker hookup.

Step 3. The speaker connection that produces the best bass response is the better connection. In other words, the speakers are operating in phase.

OSCILLOSCOPE-MICROPHONE PHASING METHOD

Test Equipment:

Oscilloscope, two crystal microphones, and an audio signal generator

Test Setup:

See Figure 3-19.

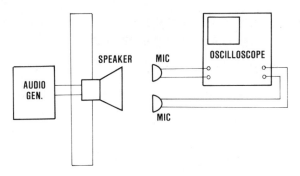

Figure 3-19:   Test setup for phasing speakers with an oscilloscope and two crystal microphones

Comments:

This test can be used to check for the correct wiring polarity of two speakers not in a complete speaker system; that is, without a crossover network.

Procedure:

Step 1. Connect one microphone to the vertical input of the scope and the other to the horizontal input.

Step 2. Place the two microphones at equal distances from one single-cone speaker. Place them an inch or so apart and just a few inches in front of the speaker.

Step 3. Connect an audio frequency signal generator to the speaker terminals and set it to operate at a low frequency ... somewhere in the vicinity of 400 Hz.

Step 4. Adjust the scope vertical and horizontal controls (you may have to adjust the signal generator output level at the same time) until you have a line slanted to the left or right on the face of the scope. Either way is okay, but make a note which way the line slants—to the left or right—because this is your reference.

Step 5. Now you're ready to test two speakers to determine if they are connected in phase or out of phase. Place one mike in front of each speaker at equal distances and use a low frequency to drive the speakers. You'll know when the speakers are wired in phase because you'll see the same slanting line that you saw in Step 4. If the line slants in the opposite direction from the one in Step 4, the speakers are out of phase and should be wired in phase.

3.13 TESTING SPEAKERS IN TWO-OR THREE-WAY DRIVE SYSTEMS FOR PROPER PHASING

Test Equipment:
Same as in Test 3.12, under the heading Oscilloscope-Microphone Method.

Test Setup:
See Figure 3-20.

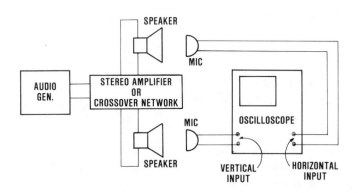

Figure 3-20: Test setup for checking the phase of speakers in a two-or three-way system using a crossover network.

Procedure:

Step 1. Connect the test equipment as shown in the block diagram in Figure 3-20 and adjust the signal generator to the crossover network frequency. If you're checking a three-way system, see Step 3 for the proper hookup.

Step 2. Set the scope for a reference pattern using the method described in the preceding test procedure.

Step 3. Place the crystal microphones a few inches out in front of the speakers you're checking. If you're checking a three-speaker system, place the mikes in front of the woofer and midrange speaker. In this case, make the polarity of the tweeter the same as the woofer, to produce the correct scope pattern; i.e., a line that slants the same way as your reference line (See Procedure in the preceding test).

Step 4. You may see a circle on the scope rather than a slanted line. For example, if the speakers under test are 90° out of phase, it will cause a circle to appear on the scope. The same rule holds true in this case. Adjust the wiring on the speakers until they are in phase. When they are in phase, you'll see the same slanted line as your reference.

Chapter 4

Practical Testing and Measuring Techniques Using Ordinary Shop Instruments

The following tests and measurements are a collection of largely ignored testing techniques that can be done with ordinary low-cost gear. Despite the literature calling for special purpose test equipment, we can still conduct successful tests on most of the electronic components found in our service shops, with nothing but an inexpensive VOM, using the oldest way known—simply measure the voltage. You'll have to make other measurements, of course, but not very often. If you know the "tricks of the trade" (Translation: Knowing the correct procedure), you can test almost anything electronic with the test gear you have in your shop right now.

The "secret", if there is one, of successful low-cost equipment testing is two-fold: One, the ability to use the test equipment you have, and two, a knowledge of a test procedure that can be applied using this test gear, to obtain the information you need. Reading the manufacturer's instruction manual for your instruments, and a little experience, will take care of number one. This chapter is aimed at successful ways of cleaning up some of the sticky areas in number two. It is written for the technician who is searching for fast, practical, and more economical ways to make tests and measurements.

To illustrate, if your VOM doesn't have a low-voltage range, simply turn to Section 4.4 for a step-by-step procedure that will quickly solve the problem. Want to measure inductance and have only a voltmeter? No problem. Turn to Section 4.8. Same thing for measuring capacitance; turn to Section 4.9. You'll also find "odd-ball" measurements such as how to measure a quartz crystal frequency using only an inexpensive dip meter, and how to check a transformer's phase with a VOM. These are just a few of the tests and measurements that are organized and coordinated for easy reference, in this chapter.

4.1   MEASURING RESISTANCE IN MEGOHMS

Test Equipment:
DC voltmeter and DC power supply. Almost any DC source will do (25 to 600 volts), all you need is a measurable voltage.

Test Setup:
See Figure 4-1.

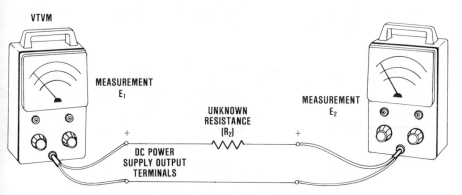

Figure 4-1: Voltmeter connections for determining the value of a resistor exceeding your ohmmeter's range

Comments:
Typically, the upper limit of direct resistance measurements with medium priced multimeters will be 2, 100, or 1,000 megohms. However, you can extend the upper limit of resistance measurements by using the DC voltmeter ranges of your multimeter, and external DC supply voltage, and the simple formula

$$R_2 = R_1 \frac{(E_1 - E_2)}{E_2}, \text{ where } R_2 \text{ is the}$$

unknown resistance value, R_1 is the DC voltmeter input resistance, and E_1 is the output voltage of the DC voltage supply, and E_2 is the voltage reading of the voltmeter when it is placed in series with the unknown resistance, as shown in Figure 4-1.

Generally, you'll find that high impedance voltmeters, such as a VTVM, will have an input resistance of 11 megohms with its DC probe attached (1 megohm in the probe). Therefore, the value of R_1 always will be 11,000,000 ohms, in this case, or you can drop the zeros and use only the number 11, which will automatically give you an answer in megohms.

Procedure:

Step 1. Connect the circuit as shown in Figure 4-1 and measure the DC power supply voltage (E_1). Let's say you measure 385 volts.

Step 2. Measure the second voltage (E_2) by connecting the voltmeter's negative lead to the negative power supply terminal, and the voltmeter's positive test lead to the point shown in Figure 4-1. Let's say that you measure 1.5 volts this time.

Step 3. Compute the value of the unknown resistance. Using the formula given in the Comments, and assuming a voltmeter input resistance of 11 megohms,

$$R_2 = R_1 \frac{E_1-E_2}{E_2} = 11 \frac{385 - 1.5}{1.5} = 2812.33 \text{ megohms.}$$

In round numbers, R_2 equals about 3,000 megohms.

4.2   VOM MOVING COIL METER TEST

Test Equipment:

1.5 volt battery and resistor (see the Procedure for correct value)

Test Setup:

See Figure 4-2.

Comments:

If you suspect the meter in a VOM is defective, do not attempt to repair it yourself. After you have determined the meter is bad, using this test, return the VOM to the manufacturer. If there is evidence of acid core solder, paste fluxes, or irons cleaned with sal ammoniac, the VOM will not be serviced by most manufacturers. Instead, the instrument will be returned unrepaired. Also, when writing the manufacturer, give a full description of the trouble, indicating which sections of the circuit you have tested. Do not try to make a continuity check of a meter movement with a standard ohmmeter because excessive current may damage the instrument.

Procedure:

Step 1. Calculate the value of current needed to provide about 2/3 full-scale deflection. For example, if the service manual shows the

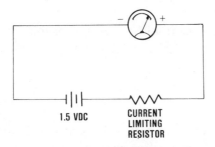

Figure 4-2: Circuit connections for checking a moving coil DC meter movement

meter movement requires a current of 39.5 microamperes to move the pointer to full-scale deflection, 2/3 full-scale current is about 26.3 microamperes (2/3 x 39.5 μA = 26.3 μA).

Step 2. Calculate the value of series resistance needed to provide 2/3 full-scale deflection using 1.5 volts DC and a current of 26.3 μA. The value of the resistor, in our example, is about,

$$R = E/I = 1.5 \text{ V}/26.3\mu\text{A} = 57 \text{ k-ohms}$$

The nearest standard value is 56 k-ohms $\pm$ 5%, which is close enough.

Step 3. Remove all connections on the back of the meter movement.

Step 4. Connect the resistor between the battery negative contact and negative terminal of the meter.

Step 5. Connect a jumper lead between the battery positive terminal and the meter positive terminal.

Step 6. When you make the last connection (Step 5), you should see a meter deflection very close to 2/3 full-deflection. If there is no deflection, or if the deflection is considerably above or below 2/3 full-scale, the meter is probably defective and should be sent in. However, it's best to go through your calculations one more time just to be sure that you are using the correct value series resistor in Step 2.

4.3   CIRCUIT LOADING TEST

Test Equipment:
  Resistor (preferably $\pm$ 1%) having a value equal to the input resistance of the instrument being checked.

Test Setup:

First, connect the test instrument leads to the circuit under test and then insert the resistor in series with the test instrument test lead (see Procedure).

Comments:

This test is good when used with high input impedance instruments such as VTVM's, FETVM's and other electronic voltmeters where circuit loading should not be evident.

Procedure:

Step 1. Connect the voltmeter across the circuit to be measured. Record the voltage reading.

Step 2. Insert the resistor in series with the voltmeter (hot lead) and again measure the voltage at the same point you measured in Step 1. Record this second voltage reading.

Step 3. The voltage you measured in Step 2 should be almost exactly one-half the voltage you measured in Step 1. If the second voltage reading is greater than one-half, it's an indication of circuit loading. In other words, your direct measurements are in error.

4.4 HOW TO USE A VOM LOW-CURRENT RANGE TO MEASURE LOW VOLTAGES

Test Equipment:

VOM and a precision resistor ($\pm$ 1%)

Test Setup:

See Figure 4-3.

Comments:

Let's assume that your VOM has an internal resistance of 10,000 ohms for 50 μA (marked 0.05 milliampere, on some meter current range settings) full-scale deflection. Using Ohm's law, we find it takes 0.5 volts to produce this much current through a 10 k-ohm resistor. Now, if the test leads of the VOM *do not* have a resistor built in, we can use this range as a voltmeter having a 0.5 volt full-scale reading. In this case, simply read the 5 volt DC scale and divide by 10, or the 10 volt DC scale and divide by 20.

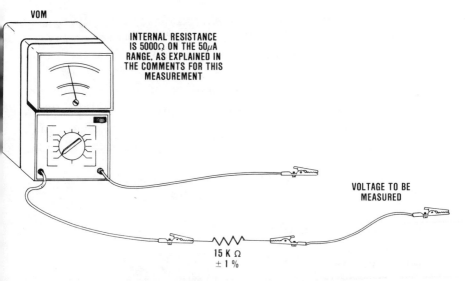

Figure 4-3: Test setup for converting VOM voltage and current ranges to measure low DC voltages

If you want to provide a 2.5 volt full-scale range, you have to increase the total resistance of the 0.05 milliamp range to 50,000 ohms. Since the meter has a total internal resistance of 10,000 ohms on the 0.05 milliampere scale, we must insert 40 k-ohms in series with the VOM's test lead. Read the 2.5 volt full-scale on the 250 DC volt scale and divide by 100.

Many VOM's, such as Heathkit model IM-105, have a voltage drop of 0.25 at full-scale, with an internal resistance of 5,000 ohms. Again, using Ohm's law, we find that 0.25 volts would have to be applied on the 50 μA range for full-scale deflection. Using this VOM as an example, if we want to provide a 1 volt full-scale reading, we must increase the total resistance of the 50 μA range up to 20 k-ohms. Now, since the meter has an internal resistance of 5 k-ohms, you need to add 15 k-ohms in series with the VOM test lead. Read the voltage on the 100 VDC scale and divide by 100.

Procedure:

Step 1. Set the VOM range switch to the 50 μA position, or any other current range you select.

Step 2. Consult the VOM operations manual to determine the internal resistance.

Step 3. Calculate the value of the series resistor needed, as explained in the Comments about this measurement.

Step 4. Insert the resistor in series with the VOM test lead and make the voltage measurement. *Remember: It's very easy to burn out a meter, so double check your calculations before you do this step!*

Step 5. Read the voltage on the appropriate DC voltage scale and divide the reading by the correction factor (see Comments).

4.5 HOW TO MEASURE THE INPUT RESISTANCE OF AN ATTENUATOR PAD

SIMPLE OHMMETER METHOD

Test Equipment:
Ohmmeter and terminating resistor

Test Setup:
See Procedure.

Comments:
There are several different type audio attenuator pads that can be checked with this procedure. Figure 4-4 shows some basic attenuator pads.

Procedure:
Step 1. Connect a terminating resistor of equal value to the pad's normal load, across the pad's output. For example, if the pad is connected between two 75 ohm lines during normal operation, use a 75 ohm resistor (the power rating of the resistor isn't important).

Step 2. Measure the input resistance of the pad with your ohmmeter. The resistance you measure should equal the normal operating impedance of the pad.

VOLTMETER AND AUDIO-SIGNAL GENERATOR METHOD

Test Equipment:
Audio frequency generator (one similar to Heathkit Model IG-1272 is excellent), high input impedance voltmeter, load resistor with a resistance value equal to the output impedance of the device under test,

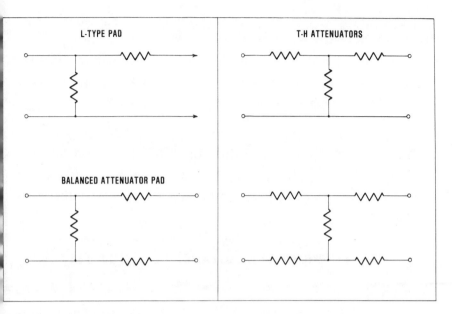

Figure 4-4: Basic attenuator pads. The H and T type attenuators generally are used only to attenuate audio signals and not for impedance matching. Therefore, their input and output impedances usually are equal.

and an adjustable precision resistor (a resistor decade box is the easiest to work with).

Test Setup:
See Figure 4-5:

Procedure:
Step 1. Connect the test setup as shown in Figure 4-5. The

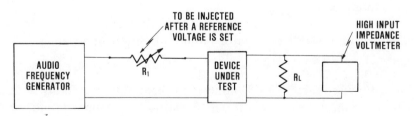

Figure 4-5: Test setup for measuring the input resistance of any resistive device having an input resistance under 10 k-ohms

variable resistor R_1 is *not* connected in series with the audio frequency generator until you reach Step 4.

Step 2. Set the audio signal generator to operate at 1,000 Hz.

Step 3. Adjust the output level of the signal generator until you have a convenient AC voltage reading on the voltmeter—about 5 or 10 volts—without R_1 in the circuit.

Step 4. Insert R_1 into the test circuit and adjust its value until the voltmeter reading is exactly one-half of what you had in Step 3.

Step 5. Measure, or read, the value of R_1. For all practical purposes, the value of R_1 is equal to the input resistance of the device under test.

4.6 ATTENUATOR REACTANCE RESPONSE TEST

Test Equipment:
Audio signal generator, voltmeter, load resistor, and variable resistor

Test Setup:
See Figure 4-5.

Procedure:
Use the same procedure explained in the last test for all frequencies of interest. At some upper frequency, the input resistance should begin to drop to a lower value due to reactance shunting. This is the point where the circuit's reactance is beginning to affect your reading; i. e., the circuit is no longer predominantly resistive.

4.7 VARIABLE ATTENUATOR PAD TEST

Test Equipment:
Ohmmeter and two load resistors having the same resistance values as the variable attenuator input and output impedance

Test Setup:
First, connect your ohmmeter to the input (or output) of the pad and connect the proper value load resistor to the opposite end. The power rating of the resistor is not important. Simply reverse the setup to make the second check.

Comments:

Many constant impedance attenuators are variable. In this case, simply adjust the step attenuator or variable control to all possible settings. The ohmmeter reading should remain essentially the same for all positions (within the tolerance of the pad). Figure 4-6 shows a simple schematic of a variable T-pad and a bridged T-pad.

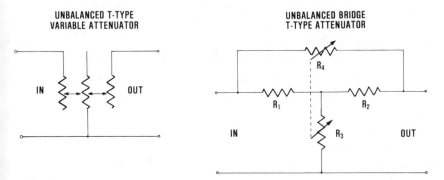

Figure 4-6: Continuously variable attenuators that can be checked with an ohmmeter

It doesn't make any difference which end of the pad you connect your ohmmeter to, to begin with (in or out). However, it is important that you connect the proper value of resistance—it must be equal to the line impedance—to the opposite end of the pad or your readings may be in error.

Procedure:

Step 1. Connect the proper value of load resistance to the output of the constant impedance variable attenuator.

Step 2. Connect your ohmmeter across the input terminals of the pad.

Step 3. Vary the attenuator control knob, or switch, through its range, at the same time watching the ohmmeter reading. The reading should remain almost constant (some pads have a little more tolerance than others, therefore you might see a slight meter variation in some cases).

Step 4. Reverse the setup and connect a load resistor having the same resistance value as the pad's input impedance.

Step 5. Connect the ohmmeter to the pad output terminals. Again, watch the ohmmeter reading as you vary the pad attenuator control. The reading should remain essentially constant. If there is little or no change in the resistance reading, the pad can be assumed to be properly matched to the lines.

4.8 INDUCTANCE MEASUREMENTS

VOLTMETER METHOD — AUDIO FREQUENCY DEVICES

Test Equipment:

High input impedance AC voltmeter, filament transformer and resistor decade box (several resistors from 50 to 10 k-ohms ± 1% tolerance can be used in place of the decade box)

Test Setup:

See Figure 4-7.

Comments:

This measurement can be used to determine the inductance of most power transformer windings or other audio frequency transformers. It also works well when applied to power supply filter chokes. *Note: Do not try to use a small transistor transformer in place of the suggested filament transformer because they are not designed to be used with a 115 VAC input.*

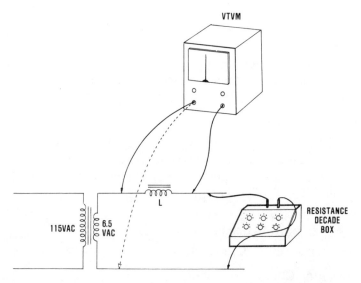

Figure 4-7: Test setup for measuring inductance with a voltmeter

Procedure:

Step 1. Measure the AC voltage across the inductor (transformer winding, choke, etc.).

Step 2. Measure the AC voltage across the resistor decade box terminals (or the resistor, if you're using a single resistor in place of the decade box).

Step 3. Either change the value of the single resistor by substituting other value resistors, or adjust the resistance decade box until the two voltages (Step 1 and Step 2) are equal.

Step 4. When the two voltage readings are the same, $X_L = R$. Therefore, all that is left to do is plug the resistance value into the formula;

$$L = \frac{R}{376.8}$$

Note: This formula is only good for a 60 Hz power line frequency. For example, let's say you have a resistance of 2,000 ohms in the circuit after you've adjusted the resistors until the two voltages are equal. Substituting 2,000 ohms in place of R, we have an inductance value of approximately 5 henrys.

AUDIO-SIGNAL GENERATOR METHOD

Test Equipment:

Audio-signal generator, AC voltmeter and a 1 ohm resistor ±1%

Test Setup:

See Figure 4-8.

Comments:

The use of an audio-signal generator greatly expands the measurement range.

Procedure:

Step 1. Connect the circuit as shown in Figure 4-8, set your voltmeter to read an AC voltage, and energize your signal generator. Set the frequency of the generator to an audio frequency that will produce a convenient reading across all components during the following steps.

Step 2. After the signal generator and components have

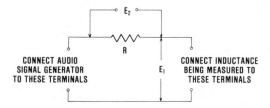

Figure 4-8: Test for measuring inductance using an audio-signal generator and AC voltmeter

warmed up for a few minutes, measure the voltage across the inductor terminals (shown as E_1 in the diagram).

Step 3. Calculate the inductance of the choke, etc., by using the formula;

$$L = \frac{E_1}{(6.28)(F)(E_2)}$$

where F is the operating frequency of the signal generator, E_1 and E_2 are explained by Figure 4-8. For example, let's assume that $E_1 = 10V$ and $E_2 = 2.65$ millivolts, when the signal generator is set to operate at 60 Hz. Our calculations are as follows:

$$L = \frac{E_1}{(6.28)(F)(E_2)} = \frac{10}{(6.28)(60)(0.00265)} = 10 \text{ henrys}$$

DIP METER METHOD—RF DEVICES

Test Equipment:
    Dip meter and one capacitor of known value. The capacitor may be any that will resonate with the inductor you are checking (see Procedure).

Test Setup:
    See Figure 4-9

Procedure:
    Step 1. Connect a small capacitor of known value and the unknown inductor in parallel, as shown in Figure 4-9.

    Step 2. Try different plug-in coils with your dip meter until you're able to measure the resonant frequency of the capacitor and coil combination. *Note:* Be careful and don't tune up on a harmonic. The second harmonic can be especially troublesome. To improve your

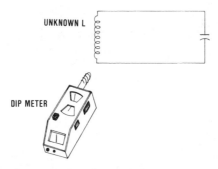

Figure 4-9: Test setup for measuring air core inductors

accuracy and help eliminate the harmonic problem, use as loose a coupling as practical. Move the dip meter as far away as possible (until the dip needle barely dips), and then tune for *maximum* dip. This should give you the fundamental frequency of the tank circuit.

Step 3. Convert the frequency reading and capacitance value to megahertz and microfarads, respectively. When you do this, it simplifies your math work and automatically gives you an answer in microhenrys.

Step 4. Use the formula:

$$L = \frac{0.025}{(F)^2(C)}$$

where F is the dip meter frequency reading in megahertz, C is the value of the known capacitor, and L is your answer in microhenrys. As an example, let's say you measure a resonant frequency of 6.9 MHz, using a capacitor of 100 pF, connected across an unknown small air core inductor. Using the formula given in Step 4, we get;

$$L = \frac{0.025}{(6.9)^2(.0001)} = \frac{25 \times 10^{-3}}{47.6 \times 10^{-4}} = 5.25 \ \mu H$$

4.9 CAPACITANCE MEASUREMENTS

SIMPLE VOLTMETER METHOD — LARGE VALUE CAPACITOR

Test Equipment:

AC voltmeter, filament transformer and resistor decade box (several resistors 50 to 50 k-ohms, or a variable 50 k-ohm resistor may be used in place of the decade box). *Note:* Do not try to use a small transistor transformer in place of the suggested filament transformer because these are not designed to operate with a 115 VAC input voltage.

Test Setup:
> See Figure 4-10

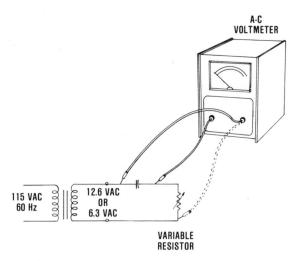

Figure 4-10: Test setup for measuring a large value capacitor

Comments:
> This method of measuring a capacitor value is not very accurate, if done with the average low-cost equipment found in many home shops. However, it is sufficiently accurate for most practical applications. The value of resistance you use will depend on how small a capacitor you're trying to measure. For example, 50 k-ohms will permit you to measure a capacitor of about $0.05\,\mu$F, at best. Any higher value resistance probably will cause considerable error in your measurement.

Procedure:
> *Step 1.* Connect the unknown capacitor and variable resistor in series with the secondary of a filament transformer, as shown in Figure 4-10.

> *Step 2.* Connect the primary leads of the filament transformer to a 60 HZ power line.

Step 3. Make alternate measurements of the AC voltage across the variable resistor and capacitor, while adjusting the variable resistor until both voltage readings are equal.

Step 4. Read the resistance value off the resistance box (or measure the resistance value with an ohmmeter, if you're using a single resistor).

Step 5. The resistance value is equal to the reactance of the capacitor when the two voltages are equal. Therefore, all that is left to do is substitute the resistance reading into the formula $C = 0.159/fR$, where C is capacitance in farads, f is frequency in cycles-per-second (Hz), and R is equal to X_C. For example, with a 60 Hz filament transformer having approximately 12 volts AC on the output, let's say your measurement across the two components is 8.7 volts. You should measure a resistance very near 12 k-ohms, in this case. Substituting these values in our formula;

$$C = 0.159/(60)(12,000) = (159 \times 10^{-3})/(72 \times 10^{4})$$
$$= 2.2 \times 10^{-7} = 0.22 \ \mu F$$

DIP METER METHOD — SMALL VALUE CAPACITORS

Test Equipment:
Dip meter and a capacitor of known value

Test Setup:
See Figure 4-11.

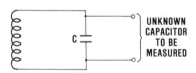

Figure 4-11: Connections for a standard resonant circuit that can be used to measure small value capacitors

Comments:
You will need a coil and a small capacitor of known value. When selecting the coil and capacitor for the standard resonant circuit, it is best to use the highest C to L ratio practical because this will produce the most accurate measurements.

Procedure:

Step 1. Select a capacitor of known value, let's say 100 picofarads. Record the value.

Step 2. Select a suitable air core inductor that will produce resonance when placed in parallel with the known capacitor. As an example, let's assume your dip meter shows the circuit to be resonant at 7.9 MHz. Record this value.

Step 3. Remove the original capacitor from the circuit and substitute the unknown capacitor.

Step 4. Again, measure the resonant frequency with your dip meter. Let's suppose you read 5.47 MHz. Record this number.

Step 5. Now, simply plug your recorded values into the formula:

$$C_x = C_1 \left(\frac{f_1}{f_2} \right)^2$$

where C_x is the unknown value of capacitance, C_1 is the value of the capacitor you selected for Step 1, f_1 is the frequency you measured in Step 2, and f_2 is the frequency you measured in Step 4. In the example,

$$C_x = (100 \times 10^{-12}) \left(\frac{7.9 \times 10^6}{(5.47 \times 10^6)} \right)^2 = (100 \times 10^{-12})(2) = \text{approx. } 200 \text{ pF}$$

4.10 MEASURING A CRYSTAL FREQUENCY

Test Equipment:
Dip meter

Test Setup:
See Figure 4-12

Comments:
It is important to remember that a quartz crystal can have more than one frequency that will produce a dip when excited by a dip meter. Generally, the greatest dip is the fundamental frequency. However, some are designed to work on overtones, especially the ones used in CB rigs. For example, the overtone oscillator of one phase-locked-loop system runs on the third overtone of 11.8066 MHz crystal; its output, therefore, is nominally 35.42 MHz. This method of checking crystals will give you the approximate frequency of the crystal. If you're

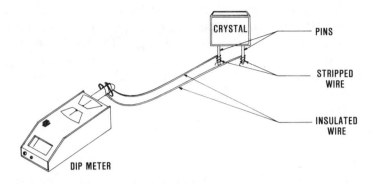

Figure 4-12: Circuit connections for checking the approximate frequency of a crystal. The meter reading of the dip meter will dip down when it is tuned to the frequency of the crystal.

checking CB crystals used in frequency synthesizers—such as a 6-4-4 synthesizer—and find one that is anywhere close to 1,000 Hz off the indicated frequency, it should be checked in a frequency counter because you can expect it to change more. Assuming your dip meter is fairly accurate, you'll probably have to replace the crystal, in this case. Most manufacturers give no accuracy figure for a dip meter. For the ones that do, it will be around ±3%.

Procedure:

Step 1. Wind a couple of turns of insulated wire around the coil of your dip meter and connect the stripped ends of the wire to the pins of the crystal, as shown in Figure 4-12.

Step 2. Plug the lowest frequency coil into the dip meter and tune for maximum dip. If you don't get a dip with this coil, try the next one up, and so on. When you find the dip, read the frequency off the dip meter. This is the approximate frequency of the crystal.

4.11 TRANSFORMER PHASE TEST

Test Equipment:
     VOM

Test Setup:
     See Figure 4-13.

Comments:
     It is possible to parallel two transformers and double the load

current that one can handle. If both transformers are identical (both secondaries have the same output voltage), you won't have problems with no-load current flow in the windings. However, if there is a difference in the output voltages, it's possible to have overheating. To check for this problem, apply power with no load attached and let them sit for about an hour or so, touching them periodically to see if they are heating. *Make this check even when the transformers are correctly phased.*

Procedure:

Step 1. Connect the primary leads as shown in Figure 4-13.

Step 2. Place a jumper lead between the transformer secondary leads 1 and 3 (See Figure 4-13).

Step 3. Measure the voltage between transformer leads 2 and 4. If the reading is zero, or very near it, the transformers are connected in phase. If you don't read zero voltage, proceed with Step 4.

Step 4. Measure the voltage of one of the transformer secondaries.

Step 5. If the voltage reading you make in Step 3 is higher (usually about double) than the voltage reading in Step 4, the transformers are connected out of phase.

Step 6. Reverse the leads of one of the transformers. Again, measure the voltage. If the meter reads zero, or very close to zero, the

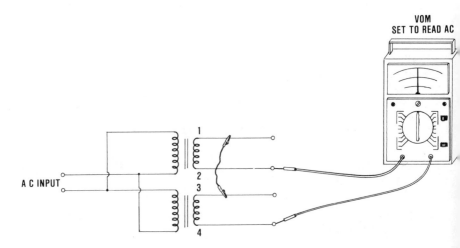

Figure 4-13: Circuit connections for checking the phase of parallel transformers

transformers are properly phased. Don't forget to check for no-load current (see Comments).

4.12   HOW TO CHECK A TRANSFORMER LOAD CAPACITY

Test Equipment:
Load resistor and thermometer

Test Setup:
See Procedure.

Procedure:
Step 1. Connect the transformer primary lead to the AC power line and let it warm up for about one hour.

Step 2. Measure the transformer temperature.

Step 3. Connect the maximum load you expect the transformer to handle, across the secondary winding. Check the temperature periodically and let it operate for about one more hour. Take another reading at the end of this hour.

Step 4. Compare the temperature you had in Step 2 to the one at the end of the hour in Step 3. You should not have a temperature change of more than 68° F (20° C) to 86° F (30° C) at any time during the test. For smaller transformers, use the 68° F temperature rise value, and on larger ones, use the 86° F value. Any temperature increase substantially below these values indicates that the transformer will run comparatively cool for your application.

Chapter 5

Tests and Measurements Needed for Maintaining Power Supplies

The successful practice of electronics, for business or pleasure, requires a knowledge of power supplies and practical testing. The DC power supply is a vital part of most electronics equipment. In fact, just plain common sense tells us that it is one of the first things to check when troubleshooting. As we all know, if there is no DC voltage, none of the other stages will work. Furthermore, the DC supply is the cause of many of our service jobs, so *it should always be tested for proper operation.* This chapter focuses its attention exclusively on the practical aspect of testing power supplies at the bench and in the field.

With just a little bit of logical reasoning, the following procedures, and a few pieces of test equipment, most DC power supplies can be tested with ease. You'll find the test gear you need listed in each test. In most cases, about the only equipment you'll need is a good AC/DC voltmeter. However, sometimes a wattmeter or scope can be a real life-saver. Although scopes have become fairly common, not everyone has a wattmeter. Therefore, Figure 5-5 shows a circuit diagram for a "home-brew" power tester that can be put together with a minimum of time, effort, and money. The setup can prove to be a very useful instrument during your first diagnosis of a trouble.

Measurement and testing techiniques are explained in detail throughout, and many shop hints are included. You'll also find the information you need to connect your instruments and equipment for each test. Finally, to make things as easy as possible, each test is categorized and cross-referenced for quick location.

5.1 DC POWER SUPPLY INTERNAL
RESISTANCE MEASUREMENT

Test Equipment:
   VOM

Test Setup:
   See Procedure.

Comments:
   A low internal resistance is most desirable because there will be very little change in output voltage when the power supply is operating under varying load conditions.

Procedure:
   Step 1. Measure the power supply output voltage with a load connected. Generally, for test purposes, the load is considered to be a pure resistance. The resistor should be a composition type and capable of dissipating the rated power output of the power supply.

   Step 2. Measure the current flowing through the load.

   Step 3. Disconnect the power supply output from its load and measure the output voltage.

   Step 4. Now, use the formula:

$$R = \frac{\text{no load voltage} - \text{full load voltage}}{\text{current reading}}$$

For example, if you measure 30 volts with no load, 27 volts with a load, and the current is 20 milliamperes, the power supply's internal resistance is 3/ 0.02, or 150 ohms.

5.2 POWER TRANSFORMER
REGULATION MEASUREMENT

Test Equipment:
   AC voltmeter and load resistor selected to draw the maximum rated transformer secondary winding current

Test Setup:
　　See Figure 5-1.

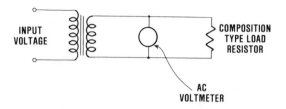

Figure 5-1: Test setup for measuring a transformer's voltage regulation

Comments:
　　It isn't unusual to find that an inexpensive transformer will not maintain a constant output voltage under varying load conditions. This is particularly important if you're replacing a transformer or designing a power supply, because it may be necessary to include extra filtering and/or regulation.

Procedure:
　　Step 1. Measure the transformer secondary output voltage without a load.

　　Step 2. Measure the transformer secondary output voltage with the load resistor attached.

　　Step 3. Calculate the transformer percentage of regulation using the formula:

$$\text{reg } \% = \frac{\text{no load voltage} - \text{full load voltage}}{\text{full-load voltage}} \times 100$$

5.3   BATTERY POWER SUPPLY CURRENT DRAIN MEASUREMENT

Test Equipment:
　　VOM

Test Setup:
　　See Figure 5-2.

Procedure:
　　Step 1. Measure the current with the transistor radio, etc., turned off.

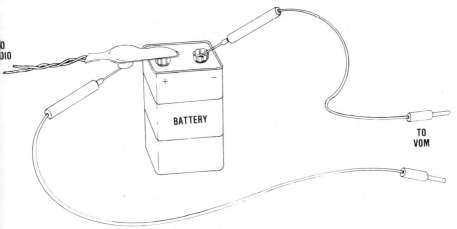

Figure 5-2: Test setup for measuring the current drain of a transistor radio battery power supply

Step 2. Measure the current with the radio turned on.

Step 3. The measurement in Step 1 should be zero. If it isn't, it's an indication of a defect in the receiver. Any current flow at all will shorten the life of the battery. When the radio is turned on, you should see approximately what the manufacturer recommends. If the current is very much higher, it's an indication that there is a fault such as a shorted transistor, etc., in the reciever. *Note:* It is not recommended that you operate a transistor radio or do this measurement while the battery is being charged. It may damage a transistor in some receivers. However, an automobile radio may be operated safely while the battery is being charged, although this can produce electrical noise in the receiver.

5.4 FIRST FILTER CHOKE CHECK
60 Hz FULL-WAVE RECTIFIER

Test Equipment:
> None

Test Setup:
> None

Procedure:
> *Step 1.* Calculate the value of minimum inductance required for the first filter choke using the formula:

minimum inductance = DC load resistance / 500

where minimum inductance is in henrys.

Step 2. Check the value of the first choke. To prevent excessive peak current, it must be equal to, or greater than, your calculated value of minimum inductance.

5.5 HOW TO MEASURE LEAKAGE CURRENT BETWEEN A POWER SUPPLY AND GROUND

Test Equipment:
High input impedance voltmeter with a millivolt full-scale range and a 1 k-ohm resistor

Test Setup:
See Figure 5-3.

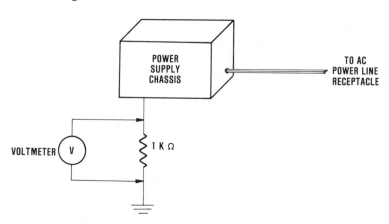

Figure 5-3: Test setup for measuring leakage current between a power supply and ground

Procedure:
Step 1. Connect the resistor between a good ground point and the power supply chassis.

Step 2. Energize the power supply and measure the voltage across the 1 k-ohm resistor.

Step 3. Calculate the current flowing through the resistor, using Ohm's law, I = E / R.

Step 4. If your answer is 3 or 4 milliamperes, it's too much. You should not have over a few microamperes.

Step 5. Reverse the power supply AC line plug and repeat Steps 3 and 4.

5.6   POWER LINE MEASUREMENT — 240 VOLTS

Test Equipment:
   VOM

Test Setup:
   See Procedure.

Comments:
   The 240 volt outlets used in homes for electric ranges, dryers and other appliances, as well as in shops, are connected to three-wire secondaries of tranformers. The schematic of the wiring and voltages is shown in Figure 5-4.

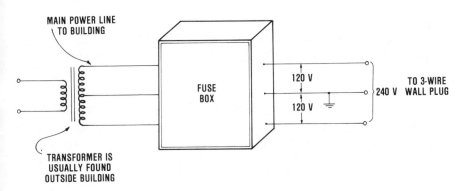

Figure 5-4: Wiring diagram for a 240 volt power system

Procedure:
   Step 1. Set the function switch to the 250 volt AC position.

   Step 2. Touch the meter probe to two of the three wall plug openings. You'll read either 240 volts or about 115 volts. If you connected between ground and one hot wire, you'll read 115 volts. Across the two hot wires, you'll read about 240 volts.

   Step 3. In some cases, you may get a voltage reading at two of the three openings. Add these readings together to get the actual value of the voltage present. The two voltages should add up to about 240 VAC.

5.7 LOW-COST INPUT POWER CHECKER

Test Equipment:
"Home-brew" test circuit shown in Figure 5-5

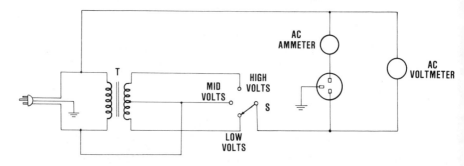

Figure 5-5: Circuit diagram for a power tester

Test Setup:
Connect the power tester between the AC power line plug and the equipment under test.

Comments:
When making the following tests, *do not* exceed the maximum current rating of the transformer. Also, the current meter must be able to withstand the maximum current the load under test will draw. Furthermore, it should be emphasized that if the equipment you're testing is reactive, your reading will be apparent power, *not true power.*

Procedure:
Step 1. Place the tester's switch (S) to the low voltage position. This will place approximately one-half the tester's center position output voltage on the load of whatever you're testing.

Step 2. Multiply the ammeter reading by the voltmeter reading. Your answer is the power consumption of the load at one-half the normal voltage input (if the load is pure resistance, it will be the real power).

Step 3. Place switch S to the middle position. Multiplying the two meter readings, gives the power at normal voltage. This is assuming you chose a transformer that produces the correct line voltage at the middle switch position to the equipment under test.

Step 4. Place switch S to the high voltage position. Multiply the two meter readings and you have the power with the equipment under test operating at 50% higher input voltage than normal.

5.8 MEASURING LOAD REGULATION — DC POWER SUPPLIES

Test Equipment:

High input impedance voltmeter and load resistor selected to draw the maximum rated current from the power supply under test

Test Setup:

See Procedure.

Procedure:

Step 1. Measure the power supply output voltage with no load.

Step 2. Connect the load resistor across the power supply output terminals and measure the voltage drop across the resistor.

Step 3. Calculate the percentage of regulation, using the formula:

$$\text{reg } (\%) = \frac{\text{no-load voltage} - \text{full-load voltage}}{\text{full-load voltage}} \times 100$$

The lower the percentage answer the better, because a low percentage means the power supply output voltage changed little when the load was connected.

Step 4. It may happen that your calculations in Step 3 will not agree with the manufacturer's specifications even though the power supply's voltage regulator is operating correctly. In this case, try using the formula:

$$\text{reg } (\%) = \frac{(10\% \text{ load voltage}) - (\text{full-load voltage})}{(10\% \text{ load voltage})} \times 100$$

It isn't unusual for the manufacturer to modify the basic equation given in Step 3, to present a better picture of what is happening under actual operating conditions. A regulated supply does not drop linearly from full load to zero load, therefore, the manufacturer does not consider the full-load range because, in the real world, we rarely reach either zero or full-load voltage.

5.9 DC POWER SUPPLY VOLTAGE TEST AND PROBABLE TROUBLE

Test Equipment:
   VOM

Test Setup:
   None

Comments:
   As with most tests, start out by checking the power supply's primary voltage supply; i. e., the AC power line.

Procedure:
   Step 1. Measure the power supply output voltage with the load disconnected. Refer to Table 5-1 if you have other than a normal reading.

CIRCUIT	LOW DC VOLTAGE OUTPUT	NO DC VOLTAGER OUTPUT
HALF-WAVE	INPUT FILTER CAPACITOR OPEN	DIODE OPEN, SURGE RESISTOR OPEN, INPUT CAPACITOR SHORTED
FULL-WAVE	INPUT FILTER CAPACITOR OPEN	DIODES OPEN OR INPUT FILTER SHORTED
HALF-WAVE DOUBLER	INPUT FILTER CAPACITOR OPEN	OPEN INPUT DOUBLER CAPACITOR, DIODE OR SURGE RESISTOR
FULL-WAVE DOUBLER	INPUT FILTER OPEN, ONE DIODE OPEN; ONE DOUBLER OPEN	FILTER CAPACITOR SHORTED
FULL-WAVE BRIDGE	INPUT FILTER CAPACITOR OPEN NOTE: ONE OPEN DIODE IN BRIDGE WILL NOT REDUCE DC VOLTAGE BY VERY MUCH, BUT WILL INCREASE RIPPLE AMPLITUDE	FILTER CAPACITOR SHORTED

Table 5-1: DC power supply testing and troubleshooting chart

Step 2. If the output voltage of the power supply is correct,connect the normal load to the supply.

Step 3. Measure the power supply output voltage. If it is up to the value shown in the schematic, fine. If it is below normal, you may have an overload somewhere in the load.

5.10 RECORD PLAYER DC POWER SUPPLY TEST

Test Equipment:
VOM

Test Setup:
See Procedure:

Comments:
The power supply transformer in small record players (both mono and stereo) is almost universally found to be the phonograph motor. A secondary winding usually is added to provide the necessary AC voltage, typically about 12 or 15 volts. It's also common to tap one of the motor windings to get the voltage needed. Figure 5-6 shows a typical record player half-wave power supply.

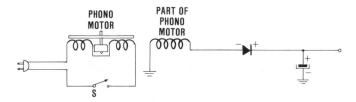

Figure 5-6: Small record player half-wave power supply using phono motor for a power transformer

Procedure:
Step 1. Turn on the record player and see if the phono motor runs slow. If it does, start feeling the motor to see if it gets hot. If it starts to get hot, turn the set off because the motor may burn out if left on long enough.

Step 2. Check the rectifier. It may be shorted.

Step 3. Measure the DC voltage across the filter capacitor. This is an electrolytic and generally the only capacitor used in the power supply.

Step 4. If you measure zero volts, check the AC input voltage with the motor running. If the AC input is normal, check the rectifier.

Step 5. If the DC voltage reads below normal, disconnect the power supply and again measure the voltage. If it's normal, you may have a short in the power supply load (generally, two or three solid state amplifiers).

Step 6. A low output voltage can be caused by an open filter capacitor. To check this, *turn power off* and bridge a new capacitor across the old one. Re-energize the system and, if the voltage returns to normal, replace the capacitor. However, as electrolytic capacitors get older they tend to develop a high leakage and, when you bridge another capacitor across them, the voltage output of the power supply will remain essentially the same. The best way to test, in this case, is to unhook the original capacitor and clip in a substitute one. This should bring the voltage back up to normal.

5.11   POWER TRANSFORMER LOAD TESTING
WITH A VARIAC

Test Equipment:
   Variac and VOM

Test Setup:
   Disconnect the power transformer and clip the Variac output leads to the rectifiers. Figure 5-7 shows an example hookup.

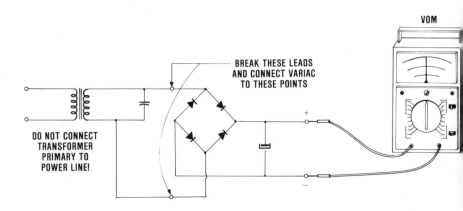

Figure 5-7: Checking a power supply transformer load with a Variac

Comments:
   Almost any power supply that does not use a center-tapped transformer, may be checked using this method. For example, a full-wave bridge, plain full-wave, or voltage doubler circuit all can be checked. Although you can get by without it, placing an ammeter in series with the Variac and rectifiers is a good safety precaution, in case there is a short in the circuit.

Procedure:

Step 1. Connect the Variac set at zero output voltage, to the rectifier input.

Step 2. Connect a VOM across the power supply DC output terminals. Set the power supply to read about mid-scale on the meter.

Step 3. Slowly start to bring the power supply up to normal output voltage. *Note:* If you do not see a slow rise in output voltage, turn the Variac back to zero output voltage *immediately* because there is a short in the circuit.

Step 4. If you do see a slow rising output voltage and it comes up to normal, let it operate for a few minutes. Assuming no overheating or other problems, the trouble is in the transformer. Check it with an ohmmeter.

Step 5. If there is no output voltage, or if it is low, check the rectifier and filter capacitor.

5.12   FILTER CIRCUIT LOAD CURRENT MEASUREMENT

Test Equipment:
VOM

Test Setup:
Break the connection between the rectifier and filter circuit. Insert the VOM, set to read DC current. Disconnect the power supply from any external load.

Procedure:

Step 1. Start by setting the VOM to its highest current reading range. The current demand may fluctuate during warm-up.

Step 2. After a few minutes of warm-up time, adjust the VOM current range switch until you have about mid-scale. Your reading is the current load on the supply, without an external load.

5.13   MEASURING THE EFFECTIVE LOAD IMPEDANCE OF A DC POWER SUPPLY

Test Equipment:
Voltmeter and ammeter

Test Setup:

Connect the DC power supply to the load under consideration, with an ammeter, all in series. Connect a DC voltmeter across the power supply output terminals.

Procedure:

Step 1. Measure the power supply output voltage with the load connected and operating.

Step 2. Measure the power supply output current with the load connected and operating.

Step 3. Calculate the effective load impedance using the formula $R = E/I$; for example, C, with a voltmeter reading of 12 volts and a current reading of 2 amperes, $R = E/I = 12/2 = 6$ ohms.

5.14   LINE VOLTAGE REGULATION MEASUREMENT — VOLTAGE REGULATED POWER SUPPLY

Test Equipment:

High input impedance voltmeter, variable load resistor, variable voltage line transformer (Variac) and ammeter

Test Setup:

See Figure 5-8.

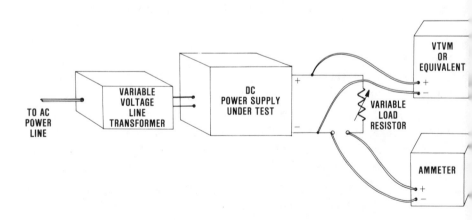

Figure 5-8: Test setup for measuring a voltage regulated power supply line voltage regulation

Comments:

Line voltage regulation frequently is measured under constant load current conditions (normally at one-half full-load current). However, better results will be obtained by measuring the regulation at both zero load and full load because, in most cases, you'll find that the power supply performs better at zero load than at full load, and this fact should be taken into consideration.

Procedure:

Step 1. Adjust the Variac for a power supply input line voltage according to the power supply manufacturer's highest operating voltage specification. If you have no specs, set it to 130 volts. Measure the DC voltage output of the power supply and note the current reading.

Step 2. Adjust the Variac for a power supply input line voltage according to the manufacturer's lowest operating voltage specification. If you don't have the specs, set it to 100 volts. Measure the DC voltage on the power supply output when the current is the same as you had in Step 1. *Adjust the load resistor to maintain a constant current reading.*

Step 3. Calculate the voltage that is half-way between the high and low settings of the Variac voltage input to the power supply. Set the Variac to this voltage and measure the power supply DC output voltage. This will be the so-called, *mid-scale voltage* in the next step.

Step 4. Calculate the voltage regulation percentage using the formula:

$$\text{line voltage regulation } (\%) =$$

$$\frac{\text{highest DC voltage - lowest DC voltage}}{\substack{\text{DC output voltage (Variac set at mid-} \\ \text{scale AC voltage input)}}} \times 100$$

5.15 SIMPLE AUTOMOBILE STORAGE BATTERY TEST

Test Equipment:
VOM

Test Setup:
Connect the voltmeter test leads across the battery terminals.

Procedure:

Step 1. Set the VOM function switch to the correct DC range and connect its test leads to the positive and negative battery terminals.

Step 2. Measure the battery voltage while the starter is cranking the engine. A six-volt system should not drop below about 5 volts. A lower voltage indicates a weak battery or other fault. A twelve-volt system should not drop below about 9 volts.

5.16 AUTOMOBILE BATTERY VOLTAGE FLUCTUATION MEASUREMENT

Test Equipment:

Transistor battery (12V), DC voltmeter with a low DC voltage range, (for example, 2.5 volts full-scale)

Test Setup:

See Procedure and Figure 5-9.

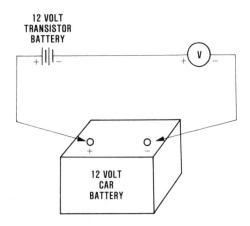

Figure 5-9: Circuit connections for measuring battery voltage fluctuations in an automotive electrical system

Comments:

It should be noted that the voltage measured on an automobile battery's terminals will not be the same when the engine is being cranked, when the ignition key is off, or when the engine is running. Furthermore, each of these conditions may have a different reading— slightly higher or lower—depending on the condition of the battery.

The following check should be made on cars that have radio transmitters which depend on their electrical systems for power.

Procedure:

Step 1. Connect the transistor 12 V battery's plus terminal to the car battery's plus terminal.

Step 2. Connect a wire from the transistor battery's negative terminal to the plus voltmeter lead, with the voltmeter range switch set to the 2.5 (or any similar setting) range position.

Step 3. Connect the negative lead of the voltmeter to the car battery's negative terminal.

Step 4. Read the voltmeter with the car engine off. You should read something like 0.5 volts, plus or minus a few tenths, if both batteries are fairly close to full charge. Ideally, you should read zero volts if both batteries are exactly 12 volts.

Step 5. Start the car engine and you should read about 2.5 volts, because the battery terminal voltage will rise to about 14 volts with the engine running.

Step 6. Adjust the car accelerator for various engine speeds and watch the voltmeter reading. If you note large changes, or any other problems such as an incorrect output voltage, it's an indication that the automobile's electrical system is not functioning properly and needs adjustment.

5.17 CONVERTING A FIXED DC SUPPLY TO A VARIABLE DC SUPPLY

Test Equipment:

Almost any fixed DC power supply and a variable power transformer (Variac)

Test Setup:

Connect a variable AC transformer in series between the AC power line and fixed DC power supply.

Procedure:

Step 1. Set the Variac to zero output voltage.

Step 2. Connect the fixed power supply to the Variac.

Step 3. Next, simply adjust the line voltage by varying the Variac until you have the desired DC output voltage from the DC power supply.

5.18   SIMPLE FILTER CAPACITOR TEST

Test Equipment:
Oscilloscope

Test Setup:
None

Comments:
The sudden appearance of multiple symptoms, AGC problems, color problems, sync instability or loss of sync, and so on, may mean you have a bad filter capacitor. The quickest and surest way to check is to use a scope. *Note:* Many instructions tell you to parallel a test capacitor across a suspected one to check for a bad filter capacitor. This is a good check, however, *never parallel a test capacitor across a DC power supply's filter capacitor in solid state equipment, with power on!* Turn the power off, parallel the capacitors, then turn the power back on to make the check. If you don't do this, the surge current may damage components such as transistors, etc.

Procedure:
Step 1. Let the scope and equipment warm up and set the scope's vertical gain to maximum deflection.

Step 2. Touch all DC supply lines with the scope's vertical test probe. In a normally operatging DC power supply replace the filter capacitor.

5.19   CHECKING A POWER SUPPLY'S RIPPLE AND OTHER PERIODIC AND RANDOM DEVIATIONS

Test Equipment:
Oscilloscope and load resistor selected to place a full load on the power supply under test

Test Setup:

See Procedure.

Comments:

Transistors and other solid state devices do not like transient spikes and surges in their power supplies. In fact, rectification spikes can destroy the logic elements in some digital systems. Sometimes you'll find all periodic and random deviations that show up on the output of a power supply, grouped together under the heading of "PARD." This stands for Periodic and Random Deviation, and the tests normally are made at 50% load current. The rectifier switching spikes will be seen on a scope as periodic vertical lines. Another deviation might be shock-excited oscillations. These will be seen on a scope as a "ringing" sine wave. Ripple usually is measured under full load power and will be seen on a scope as a sine wave (possibly, as a sawtooth), having a frequency of 60 Hz to 120 Hz. *Note:* See Comments for test 5.18 before you parallel a test capacitor across a filter capacitor.

Procedure:

Step 1. Connect the scope's vertical input probe across any of the filter capacitors except the input capacitor. You'll always find some ripple voltage on the input capacitor . . . probably about 10 volts peak-to-peak in transistor equipment power supplies.

Step 2. Adjust the scope vertical gain to maximum and set the scope to the AC mode of operation.

Step 3. Place a load resistor that will provide the desired load, on the power supply and energize the system.

Step 4. Look at the scope viewing screen. If you see any AC signal, the power supply is under suspicion. All you want to see is a simple DC line. For example, ripple levels in well-filtered power supplies should be very low. Even 0.1 volts peak-to-peak sometimes is too much.

5.20   HOW TO CONVERT RIPPLE VOLTAGE TO dB's BELOW THE MAXIMUM DC OUTPUT VOLTAGE

Test Equipment:

Voltmeter and/or oscilloscope

Test Setup:

See Procedure.

Comments:

Ideally, an unregulated DC power supply should have a ripple voltage about 80 to 100 dB below the DC voltage under full load, when operating at its rated voltage output. A regulated supply should be better—about 120 dB below the output voltage. However, these values are much better than you'll find in the average supply.

Procedure:

Step 1. Measure the maximum DC voltage output under full load. For best results, the load should be a nonreactive resistor. However, any load with little reactance, such as an amplifier, etc., will work if accuracy is not a critical factor.

Step 2. Measure the ripple voltage with an oscilloscope or other suitable instrument, such as a good quality AC voltmeter.

Step 3. Calculate the ripple voltage level below the maximum DC level using the formula:

$$dB = 20 \log_{10} \frac{\text{DC output voltage}}{\text{ripple voltage}}$$

For example, a regulated supply has 40 volts DC on its output and its ripple voltage is measured and found to be 150 microvolts rms. In this case,

$$dB = 20 \log \frac{40}{150 \ \mu V} = \text{approximately } 108$$

Chapter 6

Test and Measurement Techniques
For Tape Recorders

To get the most out of recording equipment, you must know how to maintain it; and to perform maintenance or determine if it is operating properly, you must make numerous tests and measurements. This chapter includes the testing techniques you'll need and explains how they can be performed with a minimum of inexpensive test gear.

To begin with, a recording equipment test—like all other electronics tests and measurements—requires a few preliminary checks before you start the actual test. For example, if the equipment operates from an AC line, check the voltage level first. Low line voltage can cause several problems, especially in critical phase-inversion circuits. If the recorder is battery operated, make sure that the battery is producing sufficient current to operate the machine at the recommended speeds. If there is any doubt, substitute new batteries. Next, clean and demagnetize the heads, guides, pinch roller, and capstan *before doing anything else.* Any one of these preliminary steps could clear up one or more problems such as improper tape movement, wow and flutter, or poor record and playback.

6.1 FLUTTER AND WOW TEST

Test Equipment:
New or freshly erased tape

Test Setup:
None

Comments:

Flutter in a drive system is a variation in tape speed that produces a frequency variation of the input signal at a note of 10 Hz or greater. Wow is the same as flutter except that the variation rate is from about 10 Hz down to 0.5 Hz. Below 0.5 Hz, it is called *drift*.

To accurately express the amount of flutter and wow in a percentage, generally requires an expensive flutter and wow meter. However, you can make a very good approximation of the amount of flutter by merely listening to the equipment during playback.

Procedure:

Step 1. Place a new or freshly erased tape on the recorder and record a 3,000 Hz tone. The reason for selecting 3,000 Hz is that the average human ear is most sensitive to flutter and wow at this frequency.

Step 2. Play the tape back, listening carefully. You'll hear the effects of flutter and wow as a quavering in the reproduced sound. It's interesting to note that the ear can detect instantaneous speed variations of as little as 0.1%. However, it isn't uncommon for home recorders to have instantaneous speed variations above 0.5% when they come from the manufacturer.

Step 3. Generally, if you can detect only a slight amount of flutter and wow while listening to a home recorder, it is operating properly. But, if there is enough to irritate you or it's quite noticeable, chances are that the recorder has an excessive amount. With a little practice, you can learn to judge when the machine is acceptable to the average listener.

6.2 TAPE SPEED TESTS

RULER METHOD

Test Equipment:

Any ruler that can be used to measure a certain distance on a section of recording tape and a stop watch (a wristwatch that counts in seconds may be used in place of the stop watch)

Test Setup:

None

Comments:

A linear speed check should be made on all recorders from time-to-time. Some tape recorders have speed variations of only a few seconds in a thirty-minute recording and there is no noticeable pitch change when you listen to a recording. On the other hand, you'll find there are recorders that have several minutes change in tape speed in a thirty-minute recording. With just a little experience, you'll be able to detect a pitch change even though the variation may be slight. Even if you can't hear changes in the pitch, you should check the tape speed during preventive maintenance. All you have to do is measure the tape velocity in inches-per-second using the following procedure, or use a commercially produced speed check tape if more accuracy is required.

Procedure:

Step 1. Measure off a certain length of tape and mark each end of the measured section.

Step 2. Place the tape on the machine with the beginning mark placed just before a selected point that you intend to use to start a time count.

Step 3. Start the recorder and observe how long it takes to pull the tape from the beginning mark to the end mark. Do this several times, each time noting the difference in time count. If there are more than 30 or 40 seconds difference, generally it's an indication that the tape moving mechanism is at fault. Some low-cost machines will have quite a bit of difference in time even though they are up to the manufacturer's specs. With professional machines, you shouldn't find more than a few seconds difference. Generally, eight-track recording speed is 3¾ inches-per-second and reel-to-reel is 15 inches-per-second on medium-priced machines. Incidentally, some technicians like to use a map-distance counter to make the check because it is faster and easier. If you use this system, it'll save you a little time if you'll calibrate the counter in multiples of inches-per-second before you start.

PITCH PIPE METHOD

Test Equipment:

Pitch pipe and a new or freshly erased tape. A pitch pipe is a small metal pipe that produces a fixed tone and can be purchased at most music supply stores.

Test Setup:
> None

Comments:
> As was mentioned in Test 6.1, when listening to a tone, the human ear usually can detect instantaneous speed variations of as little as 0.1%. Therefore, this method of checking a tape recorder drive system is quite accurate.

Procedure:
> *Step 1.* Place a new or freshly erased tape on the recorder.
>
> *Step 2.* Blow and record the tone of the pitch pipe.
>
> *Step 3.* Play the tape back and blow the pitch pipe at the same time. If you can detect any difference between the two frequencies, the machine is not running at the same speed as it was during the first recording (made in Step 2).

6.3 HEAD WEAR INSPECTION

Test Equipment:
> Magnifying glass

Test Setup:
> None

Comments:
> As heads wear, gaps widen. The symptoms that may be associated with head wear are a comparatively lower output level and poorer frequency response. The gap (or gaps, in stereo heads) cannot be seen with the unaided eye, by most observers. Some people say they can see them, but if you're like me, you'll need a magnifying glass, and until you have the light just right, they will still be hard to see on the first try. Also, I find that looking at a new head for comparison is the best way to identify a worn head. However, once you know what a good head should look like, you'll have no further trouble identifying a bad one. To understand why heads might wear, consider the fact that at a tape speed of only 3¾ inches-per-second, about 1.7 miles of tape passes over a head in an 8-hour period.

Procedure:

Step 1. Remove the recorder case or head cover. Be careful to note where the long and short screws are located, as you remove them. Also, the speaker and battery wires may be fairly short in battery operated cassette recorders, etc., so take care when removing the case. Incidentally, I've found that a small alligator clip works very well for holding small screws when you're replacing them.

Step 2. Use the magnifying glass and inspect the heads. The record/playback gaps should appear as a fine, uniformly wide line (or lines, on a stereo head). You'll see more than one gap (at least two) on an erase head and they will be about five times as wide as a record/playback gap. As a general rule, if the lines appear *uniformly wide*, the head is considered to be servicable. Figure 6-1 shows an example of a head used in a Panasonic cassette recorder.

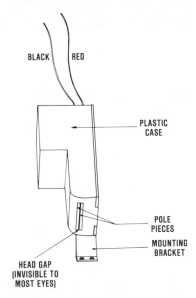

Figure 6-1: Panasonic cassette recorder head

6.4 HEAD WEAR TEST

Test Equipment:
Prussian blue (also called *Layout blue*)

Test Setup:
See Procedure.

Comments:

Prussian blue is a strong, dark blue dye commonly used to coat a piece of metal for the purpose of scribing a mark or complete layout diagram on the metal's surface. The dye usually can be found in metal-working shops and their supply stores.

Procedure:

Step 1. Coat the surface of the head where the tape makes normal contact with Prussian blue dye.

Step 2. Place a tape on the machine and run it over the head in a normal operational mode until the dye wears off.

Step 3. Remove the tape and inspect the head. You should be able to see clearly where the tape was making contact.

Step 4. If necessary, adjust the machine until the tape wipes the dye off evenly at all points of normal tape contact. If you can't adjust the head, it's generally better to replace it.

6.5 HEAD AZIMUTH TEST

Test Equipment:

Polished toothpick or any similar object

Test Setup:

None

Comments:

Excellent azimuth adjustment tapes can be obtained from several companies. These tapes have a high-frequency signal recorded on the tape at a very precise right angle from the tape's edge. If you use one of these tapes, the procedure is to play the tape and adjust the playback head for maximum output. Be careful that you are actually adjusted to maximum output because it's easy to get on a smaller peak that is just above and below the maximum output position. You can make a quick, simple test using the following procedure.

Procedure:

Step 1. Use a polished toothpick and move the tape *gently* up and down on the head while the machine is running in the playback mode.

Step 2. If you hear an increase in the output level, the head azimuth is in need of adjustment.

Step 3. If you hear a decrease in the output signal level as the tape is moved up and down, it's an indication that the azimuth is adjusted correctly.

6.6   TAPE SENSITIVITY TEST

Test Equipment:

Two or more different tapes, tape recorder, a stable audio oscillator, some form of output meter, and a tape splicer

Test Setup:

See Procedure.

Comments:

It is extremely important that recording tapes have the same sensitivity, especially when two or more tapes are spliced together. For example, consider what will happen if one tape is 6 dB less sensitive than another. There will be a sudden drop in volume when the less sensitive tape passes the playback head.

Procedure:

Step 1. Place two pieces of freshly erased tapes spliced end-to-end, on the recorder.

Step 2. Connect a stable audio oscillator to the recorder and record a 1,000 Hz signal at some level well below the distortion or overload region of the tape having the least sensitivity. Careful watch should be kept of the input signal to be sure it's held at a constant level.

Step 3. Place the tape recorder in the playback mode and monitor the output while running the tapes. If one tape has 1 dB more output than the other, that tape is 1 dB more sensitive at 1,000 Hz. This, in reality, is a measurement of a tape's frequency response at 1,000 Hz, as well as a rough check of the tape uniformity (tape coating thickness). However, it's unlikely that you'll see any change caused by non-uniformity if you are using inexpensive test equipment and checking modern recording tape.

6.7   HEAD MAGNETIZATION TEST

Test Equipment:

New or freshly erased tape and an AC voltmeter or oscilloscope

Test Setup:
   See Figure 6-2.

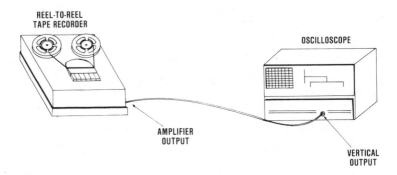

Figure 6-2: Test setup for checking head magnetization with a blank tape
              and output measuring device

Comments:

   If you hear a higher-than-normal noise when playing a tape
recording, it may be due to magnetization of one of the heads. This test
will not tell you which head is magnetized, but it will let you know if
one is.

Procedure:

   Step 1. Load the machine with new or freshly erased tape.

   Step 2. Connect an AC voltmeter or scope to the recorder
output, if the machine doesn't have a metering system.

   Step 3. Turn the input control to the off position, or short the
input circuit.

   Step 4. Run the tape for a short time (about a minute or so); just
long enough to note the reading on your monitoring device.

   Step 5. Rewind the tape and play back the section that you
passed through the recorder in Step 4 and again note the voltage
output. *Do not* change any of the control settings while performing
Steps 4 and 5.

   Step 6. Compare the two output readings and if there is any
appreciable difference, it's an indication that one of the heads is
magnetized.

   Step 7. Repeat Steps 4, 5, and 6 several times (the more, the
better; some technicians recommend ten times), looking for any

difference in readings. Any difference in readings is an indication that you should demagnetize the heads.

6.8 PRE-RECORDING TEST

Test Equipment:
New or freshly erased tape, audio-signal generator and output measuring meter

Test Setup:
See Procedure.

Comments:
It is recommended that before you record a tape, run it to another reel and use this reel as the supply reel. The reason for this is to assure yourself that the tape was not too tightly or loosely wound in some previous recording period. It also gives you an opportunity to check the tape for kinks and bad splices as well as clean the tape, if that is needed.

Procedure:
Step 1. Load the recorder with new or freshly erased tape.

Step 2. Turn the recorder input control to the off position, or short the input circuit.

Step 3. Record one or two minutes of the machine's internal noise on the tape.

Step 4. Rewind the tape so that you can play the section that was recorded in Step 3.

Step 5. Adjust the monitor speaker volume to maximum.

Step 6. With the speaker turned on full, ideally, you should not hear anything except a small amount of hiss. However, when checking inexpensive portable recorders, you'll probably hear fairly loud static. But this should be free of any erratic noises.

Step 7. If you're working with professional or semi-professional open-reel recorders, you should find almost zero noise. In this case, set the audio-signal generator to about 10 dB below the normal recording level for the machine under test. Generally, the machines are adjusted so the output indicating meter will read zero dB for peak volume.

Step 8. Record the test tone in at least three different

frequencies (low, middle, and high) and monitor the recording level during each recording.

Step 9. Play the three test tones and again monitor the output. All three test tones should be reproduced at the same level you had during recording.

6.9   BIAS ADJUSTMENT CHECK

Test Equipment:
    Audio-signal generator and output level meter

Test Setup:
    See Procedure

Comments:
    High-frequency bias current is used in all modern recorders. Generally, open-reel recorders will use frequencies of 100 kHz to 125 kHz,and modern high quality cassette recorders are about 80 kHz to 105 kHz. Since all tapes are not the same, different bias levels are required. Some home tape recorders now come equipped with separate bias switches so you can match the various bias requirements, and some recorders have no provision to change bias without changing the bias circuit component values. In the case of more expensive machines (a few hundred dollars), bias adjustment may be continuously variable as well as switchable.

    When you are checking the bias current, you may find that it is set slightly above the maximum output level, as described in the following test procedure. The amount of bias current used is determined by the individual user. When bias current is set just above the maximum current point, the distortion at low frequencies is reduced, but this curtails the frequency range of the recorder. On the other hand, when bias current is set at maximum, you'll have good frequency response. Some technicians say, "It doesn't matter because the next guy will change it anyhow." However, professional recording technicians will often chose a slight reduction of high frequency response. Furthermore, they adjust the bias to work with one brand and type of recording tape to assure optimum results at all times.

Procedure:
    Step 1. Set the audio-signal generator to a frequency between 50 and 400 Hz (probably, about 100 Hz is the best), and connect it to the recorder input terminals.

Step 2. Record the test signal and monitor the output level of the reproducing system.

Step 3. Note and mentally record the bias adjustment setting. The adjustment may be set with a screw driver or by a vernier bias adjustment on the front panel.

Step 4. Adjust the bias current, slightly. If there is a *decrease* in the output reading, adjust the other way. Watch the output. It should start increasing toward a higher reading. Continue to adjust and the reading should reach a maximum, then start to decrease. If the increase up to maximum from the original setting was very little (between ½ and 1 dB), the original setting can be assumed to be correct for optimum frequency response, with a slight correction for system instabilities. If the increase was about 1.5 dB, the original setting was correct for a comparatively low frequency response but less distortion in the low frequency range. It should be emphasized that the *best* way to adjust bias is to use the manufacturer's published data. Furthermore, you can't optimize a cassette recorder by bias adjustment alone. Generally, it's necessary to set both the bias and equalization. Sometimes the bias and equalization are separately switchable (in addition to a vernier bias adjustment). Also, some stereo machines have separate bias adjustments, and some have one adjustment that is common to both channels.

6.10 BIAS CURRENT MEASUREMENT

Test Equipment:
Oscilloscope and resistor (any value resistance from 10 to 250 ohms)

Test Setup:
See Figure 6-3.

Comments:
To prevent false readings, be sure that the recorder is connected to a ground, your scope is properly grounded, and there is no audio current present. Also, it may be impossible to change the recorder bias current without changing the bias circuit component values.

Procedure:
Step 1. Connect the resistor in the ground return lead of the record head.

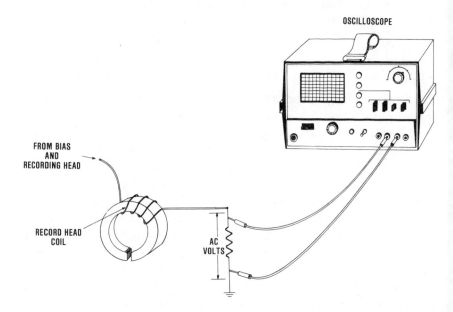

Figure 6-3: Bias current measurement circuit diagram

Step 2. Calibrate the scope to read in millivolts.

Step 3. Connect your scope across the resistor as shown in Figure 6-3.

Step 4. Measure the voltage developed across the resistor by the bias current.

Step 5. Calculate the value of the bias current by using Ohm's Law formula, I = E/R. For example, if you read 127.26 peak-to-peak, 45 millivolts rms and you're using a 10 ohm resistor, then I = E/R = 0.045/10 = 0.0045 ampere (rms). Ideally, basis should be set as high as possible without causing severe high-frequency losses in the recorded tape. However, see Comments in Bias Adjustment check (6.9) before making the final setting.

6.11 FREQUENCY RESPONSE TEST

Test Equipment:
Audio-signal generator and an external output meter, if the machine doesn't have one built in

Test Setup:
    See Procedure.

Comments:
    One advantage of the following procedure is that it automatically checks tape head alignment, frequency response of the entire system, equalizing circuits, bias oscillators, and head performance.

Procedure:
    Step 1. Using a good quality sine-wave generator, record several frequencies between 20 Hz and 20 kHz, maintaining a constant signal generator output level at each frequency you select. If the signal generator doesn't have a metered output, you can monitor the output level with an AC voltmeter *provided it maintains a constant impedance over the range of test frequencies.*

    Step 2. Attach an external VU or dB meter to the machine, if it doesn't have one built in. Then play back the tape and record the meter readings at each test frequency that you have recorded.

    Step 3. Draw a graph using the recorded readings. The vertical line should show amplitude and the horizontal, frequency. The result will be a plot of the frequency response of the recording system. Your plot should look something like the one shown in Figure 6-4.

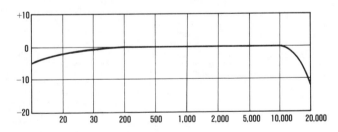

Figure 6-4: Typical frequency response chart for a recording system

6.12 DRIVE MOTOR TEST

Test Equipment:
    VOM and AC ammeter

Test Setup:
    See Procedure.

Comments:

When selecting an ammeter current range that you are going to use to check a recorder motor circuit, it's important to remember that most of these motors have very low power factors. What this boils down to is that the amount of current calculated by using the voltage and wattage rating of the motor should be about doubled for ammeter range selection. Also, motor ratings are usually given for "running" condition; the starting power will be much higher. Figure 6-5 shows a typical AC drive motor used in many older type recording machines. Figure 6-6 shows a typical AC motor and drive train used in cassette recorders.

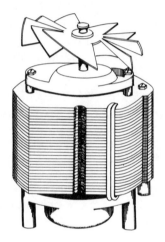

Figure 6-5: Reel-to-reel recorder drive motor

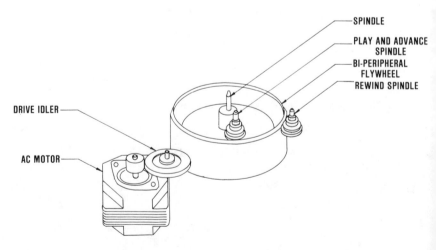

Figure 6-6: Cassette recorder AC motor and drive train.

Procedure:

Step 1. Check to see that all parts are rotating freely. On most machines, such as cassette recorders, you should be able to turn the feed, take-up reel spindles, and capstan with your fingers, with no power applied.

Step 2. Make continuity tests with your ohmmeter from the line input, through the on-off switch, to the motor coil. The resistance of the motor windings will probably be between 25 and 50 ohms.

Step 3. If you find an opening in the wiring, repair it. However, be careful you keep the motor connections in their original order. If you don't, incorrect connections may cause excessive current drain and no torque with a two-pole motor, and considerable reduction of speed and torque with a four-pole motor.

Step 4. Measure the current drain with the machine operating. On older or larger recorders, set your AC ammeter to 1 ampere full-scale, for the initial measurement. For example, if the motor consumes 60 watts, using 115 VAC, this works out to be about ½ ampere for a properly operating machine. Twice this value is nearly 1 ampere. Or, a small cassette recorder that consumes 5 watts using 120 VAC will draw 0.0416 amperes (about 42 milliampere). Doubling this, we have slightly over 80 milliamperes.

Step 5. If the motor is turning too fast, or slow, the machine may have a transistorized speed control system (see Figure 6-7). Frequently, you'll find an adjustable resistor in the speed control

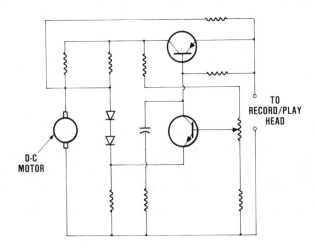

Figure 6-7: Transistorized speed control of a DC drive motor

circuit. Use a speed check tape and try adjusting the speed. Also, it may be a case of binding, due to dirt in the drive system. Use the manufacturer's recommended cleaning and lubrication procedures and materials. If the transistorized speed control won't clear up the problem, you'll probably have to replace the motor.

6.13   LOW-COST HEAD DEMAGNETIZER (DEGAUSSER)

Test Equipment:
Soldering gun and about six feet of copper wire

Test Setup:
See Procedure.

Comments:
A recording or reproducing head can become magnetized due to (1) working near the head with magnetized tools, (2) removing circuit components (such as tube or the record head) while the machine is in the record mode, (3) using an ohmmeter to check the head, and (4) rapidly changing from play to record. Although erase heads normally don't need degaussing, they can become magnetized due to a faulty bias oscillator. Therefore, a head demagnetizer is a useful tool for all heads and other parts of all magnetic recorders. The following procedure gives construction details for a simple degausser that requires no more than a few feet of wire and a soldering gun.

Procedure:
Step 1. Obtain a piece of copper wire, gauge 8, 10, 12, or 14, about six feet long. The wire may be insulated or bare, single-strand. If it is bare, it's better to insulate it.

Step 2. Wrap the wire around any type round rod that will produce about 20 turns, using the smaller wire, and about 12 turns using the large wire. Slip the wire off the coil form.

Step 3. Remove the regular soldering tip from a soldering gun. The soldering gun can be any wattage between 140 and 250 watts.

Step 4. Connect the coil to the soldering gun, as shown in Figure 6-8.

Step 5. To use the degausser, first disconnect the tape recorder from its power source. Next, place the degaussing coil over the parts to be demagnetized and energize the soldering gun.

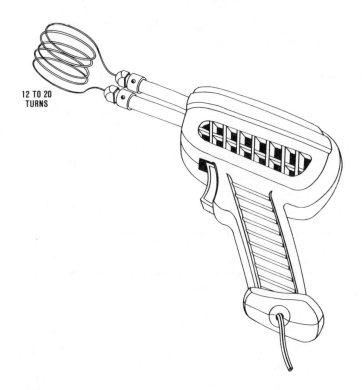

12 TO 20
TURNS

Figure 6-8: Coil connections to a soldering gun for a degaussing tool

Step 6. The time any degausser should be left near a head surface varies with the strength of the degausser's magnetic field and can be determined by a series of simple test runs. However, watch it! A degausser can magnetize just as well as it demagnetizes. For example, place a metal screw driver in the coil, energize the soldering gun for just a few seconds and you'll probably find that the screw driver is magnetized. On the other hand, bring the coil up to the head (or screw driver) slowly, then remove the coil from the vicinity of the head very slowly (taking several seconds to pull it completely away), turn it at right angles to the head and de-energize the soldering gun, and you should remove all magnetism.

6.14   LOW-COST TAPE ERASER

Test Equipment:
Permanent magnet speaker with the magnet exposed and adhesive plastic tape (for example, Scotch tape)

Test Setup:

See Procedure.

Comments:

This method of erasing tape is *not recommended* in applications where precision recording is of prime importance.

Procedure:

Step 1. Wrap adhesive plastic tape around a speaker's permanent magnet. Obtain a finish as smooth as possible because the tape must pass over the finished product.

Step 2. Place the tape to be erased on the machine and then thread it around the speaker's permanent magnet, as shown in Figure 6-9.

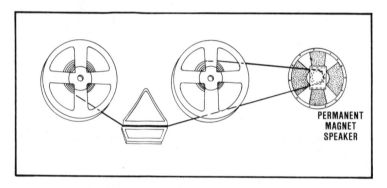

Figure 6-9: Method of threading a magnetic tape on a permanent magnet speaker to erase the tape

Step 3. Move the tape recorder take-up spool by hand to see if the tape is passing over the speaker magnet smoothly. If everything is okay, set the tape recorder to fast forward (or fast rewind, depending on which way you need to go) and, as the tape passes over the speaker magnet, it will be erased. Of course, this method is not as good as a bulk eraser, but in many applications, it will prove satisfactory.

Chapter 7

Practical TV Receiver Test and Measurement Guide

This chapter is very comprehensive, but stops short of requiring expensive professional-type equipment. Each test presents techniques that can be performed by a serviceman, technician, or experimenter, with a minimum of low-cost test gear.

The VOM and its offsprings, such as the FETVM, are unquestionably the most popular and useful service instruments found in any shop. Therefore, you'll find that the majority of the tests included in this chapter use no more than a VOM and associated probes. Of course, some of the tests require the use of instruments such as a signal generator and VTVM, for instance. However, even these use the least expensive instruments available. On the other hand, some of the tests can be made without any instruments, and others need only a couple of electronic components that, more than likely, you already have in your spare parts box.

All test setups, procedures, and results obtainable are described or illustrated in detail. You'll also find just about every practical test and measurement you'll need to check out a TV receiver is categorized and referenced in the Table of Contents in the front of this book.

7.1 PICTURE TUBE VIDEO SIGNAL TEST

Test Equipment:
   DC voltmeter and demodulator probe

Test Setup:

Remove the picture tube sockets and connect the voltmeter test probe to the signal input pin receptacle (the pin hole) in the socket. Then connect the voltmeter ground lead to a TV chassis ground point.

Comments:

The voltage reading you should expect will vary from set-to-set and also depend on the type multimeter you use (VOM, VTVM, etc.). However, if the set is operating properly, your reading will probably be between 10 and 40 volts. For safety's sake, set your range switch to the 100 volt position to start, and then decrease it for an upper-scale reading, for best accuracy.

Procedure:

Step 1. With the TV set turned off, remove the back cover and disconnect the picture tube socket.

Step 2. Connect your VOM between the picture tube socket signal input pin hole and chassis ground. As a general rule, the color code for hookup wire is green for signal carrying wire and black for ground. *But not always,* so proceed with caution when using this premise.

Step 3. Energize the TV receiver and tune it to an operating channel. In many parts of the country, you can receive a test pattern on some channels. In most locations, you'll find it broadcast in the early morning or late at night (usually before 6 AM, or after midnight). Generally, you'll get the most dependable results using a test pattern as a signal input during testing, particularly if you're comparing your readings to service notes. Of course, you can use a pattern generator if you happen to have one.

Step 4. If the receiver is operating normally, you should read a signal voltage of about 25 volts, ± 5 or 6 volts, at the picture tube socket.

7.2 RATIO DETECTOR OUTPUT SIGNAL TEST

Test Equipment:

High input impedance DC voltmeter with a demodulator probe

Test Setup:

See Figure 7-1.

Comments:

A below normal reading may be caused by an improperly tuned system and the alignment should be checked if all other circuit components are found to be good (see Chapter 1 for a method to check the diodes with an ohmmeter). *Note: Do not try to do any aligning without proper equipment and instructions.*

Procedure:

Step 1. Connect the DC voltmeter demodulator probe across points 1 and 2 (these points will require a sensitive voltmeter), or points 3 and 4.

Step 2. Observe the voltmeter reading with the set tuned to a station transmitting a test pattern, if practical.

Step 3. If you're using a multimeter, such as a VTVM, with a full-wave demodulator probe, you should see a reading that compares favorably with the manufacturer's service notes. Other instruments, such as a VOM, will probably produce readings that are somewhat lower than the manufacturer's specs. However, if you know what it should be under these conditions, a VOM can be used in place of a more expensive instrument.

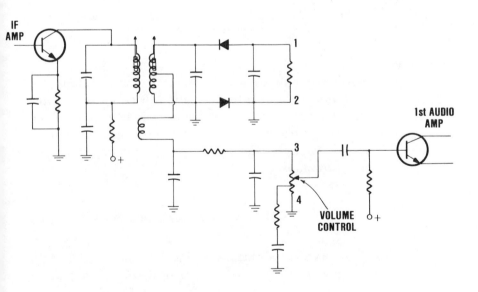

Figure 7-1: Typical solid state ratio detector circuit showing connection points for a demodulator probe

7.3   LIMITER-DISCRIMINATOR OUTPUT SIGNAL TEST

Test Equipment:
    High input impedance DC voltmeter with a demodulator probe

Test Setup:
    See Figure 7-2.

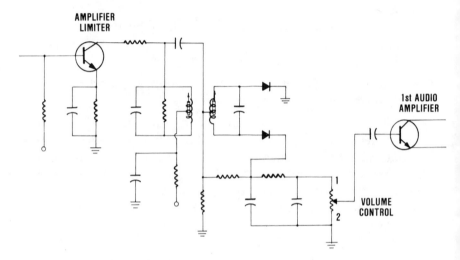

Figure 7-2: Example of a solid state limiter-discriminator showing connection points for a demodulator probe

Comments:
    The discriminator shown in Figure 7-2 is merely an example of circuit design. In practice, you'll find variations between sets. A low voltage reading at points 1 and 2, in Figure 7-2, may result from improper alignment or trouble in the circuits preceding the limiter-discriminator. *Do not try to do any aligning without the proper equipment and instructions.*

Procedure:
    Step 1. Connect the DC voltmeter demodulator probe across points 1 and 2.

    Step 2. Observe the voltmeter reading with the set tuned to a station transmitting a test pattern, if practical (or use a pattern generator).

    Step 3. Compare your voltage reading with the receiver service notes, or those obtainable from a receiver of the same type, in good operating condition.

7.4   VIDEO DETECTOR TEST

Test Equipment:

DC voltmeter and high impedance probe

Test Setup:

Connect the voltmeter probe across the video detector load resistor. Figure 7-3 shows an example of a diode detector and associated circuits.

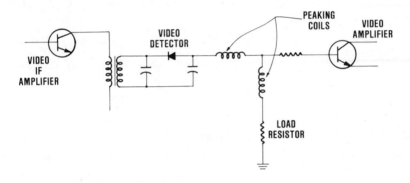

Figure 7-3: An example of a video detector system

Comments:

Generally, you'll have little or no trouble with the video detector in a TV set because there are no high voltages to contend with. However, if your voltage reading in the following procedure is zero, the first thing to do is check the diode (see Chapter 1 for a diode checking procedure).

Procedure:

Step 1. Connect the high impedance test probe of a VOM across the video detector load resistor.

Step 2. Tune the TV receiver to a station that is broadcasting a test signal, if practical (a pattern generator can be used).

Step 3. Observe the DC voltage reading. If you're tuned to a test pattern, the voltage reading should be steady. On the other hand, if you're tuned to a station broadcasting regular programming, you'll see the voltage reading fluctuating. In either case, a zero, or very low DC voltage reading, indicates a trouble in the video detector or some preceding stage.

7.5 VIDEO IF STAGE TESTS

HOW TO CHECK AN IF STAGE WITHOUT INSTRUMENTS

Test Equipment:

A 2 mfd, 50 working DC volts capacitor for checking solid state circuits, or a 0.02 mfd, 600 working DC volts capacitor for checking tube circuits, two alligator clips and a piece of hookup wire about a foot or so in length

Test Setup:

See Procedure.

Comments:

A dead IF stage can be found very quickly using the following procedure. Another use for the test setup is as a "field fix." For example, suppose an IF amplifier tube becomes defective during your favorite TV program. No problem. After pulling the tube, simply bridge the tube by placing a 0.02 mfd capacitor between the grid and plate connection. Naturally, the audio and picture will not be as good, but you may be able to see the end of the program, or get by for a few days. To make this test, or bypass an IF stage, you'll need a capacitor test lead, as shown in Figure 7-4.

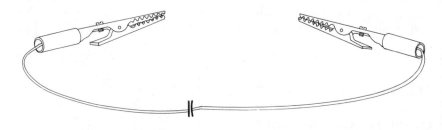

Figure 7-4: Capacitor test lead that can be used to check or substitute for an IF stage

Procedure:

Step 1. With the TV receiver off, pull the last IF amplifier tube and connect one of the alligator clips of the capacitor test lead to the input of the IF stage and the other to the output.

Step 2. Turn the TV set on. If this stage is "dead," you should see a picture and hear sound—not as good as they were, but they should come in.

Step 3. If the sound and/or picture does not return, the last IF probably isn't defective. In this case, use the same method to check the other IF stages.

RF SIGNAL GENERATOR METHOD

Test Equipment:
RF signal generator modulated by a 400 Hz signal

Test Setup:
See Procedure.

Comments:
The signal generator can be any low-cost single-signal type capable of generating signals having frequencies up to the IF frequencies of the receiver under test.

Procedure:
Step 1. Set the signal generator to the IF frequency of the stage being tested. If, as is often the case, the IF's are stagger-tuned, the input circuit is not tuned to the same frequency as the output circuit. Figure 7-5 shows typical stagger-tuned IF frequencies.

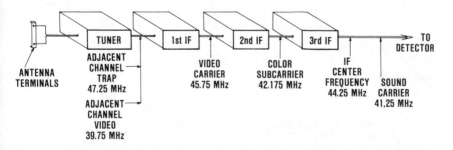

Figure 7-5: Typical stagger-tuned IF frequencies

Step 2. Connect the signal generator output leads to the input of the last IF stage. With the TV receiver operating, the 400 Hz modulation of the signal generator will produce sound bars on the screen if all stages after—and including—the last IF are working.

Step 3. Inject the signal into the input of each IF stage. When the signal generator fails to produce sound bars on the screen of the picture tube, you will have localized the defective stage.

7.6   IF SYSTEM NOISE LEVEL TEST

Test Equipment:
VOM

Test Setup:
Connect the VOM leads across the TV receiver video detector load resistor. See Figure 7-3 for an example video detector system showing the load resistor. Also, short the TV receiver antenna terminals.

Comments:
Higher-than-recommended DC voltage readings across the video detector load resistor indicate that the receiver will perform poorly in a weak TV signal area.

Procedure:
Step 1. Short the TV receiver antenna terminals.

Step 2. Connect your VOM across the set's video detector load resistor and observe the DC voltage reading, while the receiver is operating.

Step 3. Compare your reading with the DC voltage reading taken from a TV set that is known to be in good operating condition. Or, consult the manufacturer's service notes for the recommended voltage reading. In any case, the lower the reading, the better.

7.7   VHF RECEIVER FRONT-END TEST

Test Equipment:
Any signal generator capable of tuning 55.25 MHz and that can be modulated with an audio signal either internally or externally. The signal generator's frequency accuracy is not of prime importance.

Test Setup:
>   See Procedure.

Procedure:
>   *Step 1.* With the TV receiver turned on, set the receiver tuner to Channel 2.

>   *Step 2.* Set the signal generator to a frequency of about 55.25 MHz and apply an audio-modulated signal to the generator. It is very difficult—if not impossible— to accurately read the frequency setting on many low-cost signal generators. Therefore, you'll probably have to vary the frequency setting of the generator around 55 MHz to find the carrier 55.25 MHz. How to tell when you have it correct is explained in the following steps.

>   *Step 3.* Connect the signal generator test lead to the input of the TV receiver mixer. Now, tune your signal generator slightly below and above 55 MHz and watch for sound bars on the receiver viewing screen. The number of sound bars you see will depend upon what audio frequency you're using to modulate the signal generator. You may have to vary the fine tuning control to get the sound bars. However, if sound bars appear, the mixer stage is operating.

>   *Step 4.* Connect the signal generator leads to the RF amplifier input and repeat the procedure given in Step 3. If you again get sound bars, this circuit is working.

>   *Step 5.* Change the signal generator frequency to the receiver's Channel 2 local oscillator frequency. Remember, in most cases, the local oscillator frequency is the sum of the incoming station frequency and the receiver's intermediate frequency. Also, don't use the audio modulation in this, or the next, step.

>   *Step 6.* Connect the signal generator in place of the local oscillator. If the TV receiver returns to normal operation (even for a second), it is an indication that the local oscillator is probably defective—assuming Steps 3 and 4 were found to be in working order.

7.8   AGC VOLTAGE TESTS

AGC CIRCUIT TEST

Test Equipment:
>   VOM

Test Setup:
Connect the VOM test leads between the AGC line and chassis ground.

Comments:
The TV receiver service notes are not necessary (however, they *are* helpful) to make general checks of AGC operation because, as you tune the TV receiver tuner, you should see voltage variations on the VOM, or the AGC system is not functioning.

Procedure:
Step 1. Set your VOM function switch to the DC position and connect it between the AGC line and chassis ground.

Step 2. Turn the TV receiver on and tune it to a no-station position on the tuner. Take note of the DC voltage reading.

Step 3. Next, tune the TV receiver to the strongest TV station in your area and, again, watch the voltmeter reading.

Step 4. Switch the TV receiver tuner back and forth between no-station and strong signal positions and you should see voltage variations occur. If you don't, the AGC system is not operating properly. Your next step is to check transistor and component parts.

AGC PROBLEM AT THE RF AMPLIFIER TEST

Test Equipment:
None

Test Setup:
None

Comments:
An AGC trouble in the RF amplifier will cause snow on the strong channels and the weaker channels may function properly.

Procedure:
Step 1. Disconnect the AGC line.

Step 2. Energize the TV receiver and tune it to a strong station.

Step 3. If the snow disappears, you have a problem in the AGC circuit.

Step 4. If the snow does not disappear (try all channels), check the first RF amplifier because, more than likely, it is causing the trouble.

7.9   SIMPLE VERTICAL LINEARITY TEST

Test Equipment:
   None

Test Setup:
   None

Procedure:
   Step 1. Turn the receiver on and tune it to a station.

   Step 2. Misadjust the vertical hold until you see the picture slowly start to roll (lose sync).

   Step 3. As the picture rolls, you'll see a black blanking bar across the screen. By careful adjustment, you can get the blanking bar to hold fairly stable at the top and bottom of the screen.

   Step 4. Check the thickness of the bar. It should have the same thickness at the bottom of the screen as at the top. If you see any variations in thickness, linearity of the set is off.

7.10   CHECKING FOR SYNC PULSES

Test Equipment:
   A 2 mfd, 50 working DC volts capacitor for checking solid state systems, or a 0.02 mfd, 600 working DC volts capacitor for checking tube-type sets, one alligator clip, a test probe, and about a foot of hookup wire

Test Setup:
   See Procedure.

Comments:
   To make this test, you'll need to construct a capacitor test lead. All that is required is shown in Figure 7-6. *One note of warning:* When connecting to different points, be careful of strong signals. For example, output amplifiers may develop signals that are strong enough to damage the audio stages used in this procedure. In fact, it's better not to connect to any output amplifier when working on solid state chassis.

Figure 7-6: Capacitor test lead that can be used to check for the presence of sync pulses

Procedure:

Step 1. With the TV receiver off, connect the alligator clip of the capacitor test lead to the input of the audio output stage. Figure 7-7 shows an example of the connection in the sound section of a solid state color receiver.

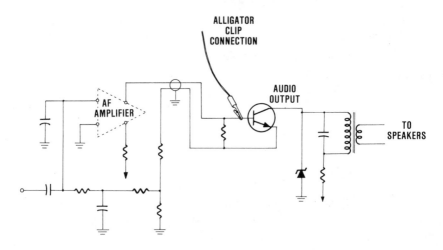

Figure 7-7: Test connections for using the sound section of a solid state color receiver for a signal tracer

Step 2. Turn the volume control to minimum.

Step 3. With the TV receiver operating, touch the test probe end of the capacitor test lead to the input, or some midpoint, of the sync section. An example of a connection point in a solid state TV would be the predriver or driver transistor.

Step 4. If the sync signal is present, you should hear a buzz in the speaker.

7.11 SIMPLIFIED HIGH VOLTAGE CHECK

Test Equipment:
Neon lamp voltage checker

Test Setup:
None

Comments:
If you simply want to check to see if there is a high voltage output, all you need is a number 51 neon lamp placed on the end of a wooden stick (dowel rod or old broom handle) about 10 inches long. See Figure 7-8.

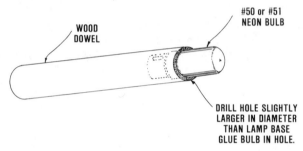

WOOD
DOWEL

#50 or #51
NEON BULB

DRILL HOLE SLIGHTLY
LARGER IN DIAMETER
THAN LAMP BASE
GLUE BULB IN HOLE.

Figure 7-8: Construction details for a homemade high voltage checker

Procedure:
Step 1. To use this simple high voltage probe, place the neon bulb near a high voltage point (for example, the cap of a horizontal output tube).

Step 2. If there is a high voltage present, you'll see the bulb glow. Another thing you'll notice is that the closer the bulb is to the high voltage point, the brighter it will glow. With just a little practice, you can learn to make a pretty fair approximation of the value of any high voltage. For example, oscilloscope cathode ray tubes, closed circuit TV picture tubes, and automobile spark plugs, are just a few you can check with this tester.

7.12 TV BOOST VOLTAGE MEASUREMENT

Test Equipment:
VTVM and a special high voltage DC probe that provides 100 times attenuation

Test Setup:
None

Comments:

Checking TV boost voltages can be a problem. Normal values of 1 KV often have a superimposed peak pulse of 1200 volts. A commonly occurring pulse form in the flyback circuit has a 15 kHz repetition rate and about 10 μ sec duration; sufficiently long enough to destroy your meter. The boosted B voltage in TV receivers is the voltage resulting from a combination of the B plus voltage from the power supply and the average value of voltage pulses (the 15 kHz, 10 μ sec pulses) coming through the damper tube from the horizontal deflection coil circuit. Although these pulses are smoothed by filtering, they are normally several hundred volts higher than the B plus voltage. One reason for the high voltage DC probe is to protect your VOM against these AC pulses. Another reason is that the length of the probe provides some protection for the user. *Warning: When servicing these probes, avoid handling the internal probe resistor! It is very susceptible to damage by moisture.*

Procedure:

Step 1. Turn the TV set on and attach the probe to the voltmeter. There are some special voltmeters mounted on probes that are built especially for this test which do not require the use of a VTVM. However, any good voltmeter with a high voltage probe can produce an accurate reading.

Step 2. Apply the probe between the point of measurement and the chassis ground (such as the high voltage connection on the picture tube, the plate of the damper tube, or plate of the horizontal output tube).

Step 3. Turn the brightness all the way down and note the high voltage reading. Remember, you're using a 100 to 1 attenuation factor, therefore, you must multiply the scale reading by 100 to determine the value of the DC voltage at the point under test.

Step 4. Turn the brightness all the way up and note the high voltage reading (again multiplying by 100, if you want to know the actual DC voltage at the test point).

Step 5. Compare the two readings and if there is considerable difference, it means the system is not operating properly.

7.13 LOW-VOLTAGE POWER SUPPLY TEST

Test Equipment:
 VOM

Test Setup:
 See Procedure.

Comments:
 The low voltage supply in a color TV receiver is similar to the low voltage supply used in black and white receivers, in that it furnishes the B plus and filament voltages (even all-solid state have picture tube filaments). However, a color receiver requires more power than a monochrome receiver. Therefore, you'll find a greater number of stages in these supplies. In fact, it isn't unusual to find 5 or 8 different value DC voltages in the output circuits of a color set's low power supply.

 When you're making tests in solid state TV receivers, it's important to remember that the operation of transistors and other semiconductor devices can be seriously affected by slight voltage changes. The power supply voltage must be fairly constant and ripple free. Therefore, power supplies in solid state TV receivers generally include a voltage regulator circuit.

Procedure:
 Step 1. Measure the voltage across the output filter capacitor. Some examples of what you should read are: 410 VDC on an RCA CTC 16 chassis (this is the output voltage on the bridge circuit), 210 VDC across one capacitor in a GE chassis CA, and 128 VDC across the final filter capacitor of one of Zenith's solid state sets. If the voltage you measure is not up to the manufacturer's specs, check the diodes and filter capacitor (or capacitors).

 Step 2. If the diodes and filter capacitors check out all right, test the power transformer secondary for possible shorted turns. You can do this with your VOM set to read resistance. Be sure to check between each half of the secondary (if the secondary is center-tapped). If there is an appreciable difference in readings, you probably have a partial, or total, short. *Note: Don't forget to pull the AC plug and bleed those capacitors **before** making resistance measurements!*

 Step 3. If the tests just described do not localize the trouble, check all other components and circuits such as the voltage regulator circuit, switches (there may be several switches, in some chassis), thermistor, and circuit breaker. You should measure about 1 volt across a thermistor after a short interval of operation (or measure approximately 120 ohms, when it is cold). Of course, you should measure zero voltage across circuit switches and circuit breakers when the set is opeating. A typical voltage measurement on the base of a regulator transistor (or the regulator output), should be about 24 VDC. However, as always, either check a set of the same make or refer to the set's schematic.

7.14 LOW-COST PICTURE TUBE TESTER

Test Equipment:
  Picture tube brightener

Test Setup:
  See Procedure.

Comments:
  The useful life of a TV picture tube usually is determined by the weakening of the electron emission in the beam from the oxide-coated cathode. Therefore, a safe, quick, and inexpensive way to check a picture tube that appears to be at the end of its life is to connect a picture tube brightener in the set. This will increase the power applied to the heater and raise the temperature of the cathode in the emission depleted tube.

Procedure:
  Step 1. Turn the set on, then darken the room. Even if you can only get a *very, very, faint,* glimmer of raster, the picture tube life can usually be extended. *Even if you don't see any light, continue this test.*

  Step 2. Turn the TV receiver off and connect a TV tube brightener between the picture tube and its base socket. A picture tube brightener for your set can be purchased at almost any electronic's supply store such as Radio Shack and, in most cases, comes with instructions.

  Step 3. In many cases, the picture will snap in (sometimes, from absolutely nothing on the screen). If it does, the life of the picture tube may be extended by the simple addition of a brightener. TV picture tube brighteners for several of the most popular brands of TV sets in your area are handy things to have in your shop if you plan on servicing TV sets. Another trick is to place two brighteners in series to boost the filaments up another few volts (3 or 4 additional volts) for a short period of time, to help rejuvenate the tube.

7.15 SIMPLE PICTURE TUBE FILAMENT TESTER

Test Equipment:
Pilot lamp number 44 or 46, blue bead

Test Setup:
Remove the picture tube socket and insert the test lamp prods into the filament connections. See Comments.

Comments:
Sometimes, it is difficult to determine if the picture tube filaments are open, or whether it's just a bad socket connection. A quick, inexpensive filament checker can be made by soldering two insulated wires to a pilot lamp, as shown in Figure 7-9.

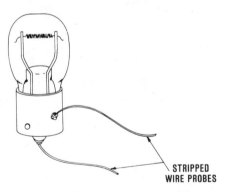

STRIPPED
WIRE PROBES

Figure 7-9: Low-cost TV picture tube filament checker. Although this illustration shows the wire connected directly to the lamp, a suitable lamp socket may be more convenient.

Procedure:
Step 1. Remove the picture tube socket and insert the test lamp leads into the filament connections. Connect them either way. It makes no difference because the filament voltage is usually an AC voltage (about 6 or 7 volts).

Step 2. If the lamp lights, the picture tube probably has burned out filaments. If not, the socket probably is the source of the trouble. This simple, little tester can save you a lot of time and trouble.

7.16 HOW TO REMOVE CRT FACE PLATE BURNS

Test Equipment:
   None

Test Setup:
   See Procedure.

Comments:
   Older TV receivers and other video display devices that use phosphor coating on the kinescope, are often found with burned spots on the viewing screen. This also can happen on color TV receivers, if a color bar generator is left on for too long a period of time with the brightness and color controls set to a high level. To prevent this from happening when testing a set, it is recommended that you don't leave a color bar generator on a set for over ten minutes or so, and the controls should be set for a fairly low intensity throughout most of the testing period. However, if there are burns, use the following procedure and it might save the day.

Procedure:
   Step 1. Tune the TV set to a strong black and white signal.

   Step 2. Adjust the vertical control until the picture rolls continuously.

   Step 3. Turn up the brightness and contrast controls for full intensity.

   Step 4. Leave the TV set running in this manner for about 3 to 6 hours. Normally, you'll see the burned spots disappear (or almost disappear) in a few hours.

Chapter 8

Practical CB Radio Tests
And Measurements

With over 25 million CB radios in use and service needs growing every day, this is an area that you may find profitable, and it looks promising for years to come. With this chapter, and a little practice, you probably can accomplish a complete system "check-out" in not much more time than it takes to read each individual test or measurement needed. You'll find this chapter explains a variety of CB tests and measurements that can be performed by any electronics technician *without an FCC First or Second Class license.*

While it's true that many electronics experimenters/CB operators do not have extensive electronics shop facilities, this should not deter them from enjoying the benefits of the many tests which require no more than simple, homemade test gear (in fact, in many tests, no test equipment at all is needed) found in the following pages. CB system checkout ... from both a technical and cost point-of-view ... is simple. You can do it yourself.

Although you can service CB equipment with a nonradiating dummy antenna connected to a transceiver, it is also *important* that you realize that if any *internal adjustments* are made on the set, it must be checked out by a person holding a valid First or Second Class FCC license before it is operated using a radiating antenna. Follow these rules and you can't go wrong:

(1) Do not attempt to operate any transceiver until a suitable dummy antenna is connected to the set (assuming you have it in a shop for testing).

(2) Don't operate the CB unit for extended lengths of time until you have checked the standing wave ratio (SWR). You can tolerate an

159

SWR up to 3 to 1 for short periods of time. However, the output power will be greatly reduced, and it's possible to damage the unit with an SWR greater than 3 to 1.

(3) Don't test the system on the air without complying with FCC rules and regulations. FCC regulations strictly limit the amount of "on-the-air" testing that may be done and, as previously explained, after anything is done to a set that affects frequency, modulation, or power (in other words, any internal adjustments or repairs), your work must be checked by a technician holding a valid First or Second Class FCC license.

8.1   TIME-SAVING ANTENNA TEST

Test Equipment:

12 volt car lamp (incidentally, a number 1815 or 1487 twelve-volt pilot lamp will also work) and screwdriver or jumper lead

Test Setup:

See Figure 8-1.

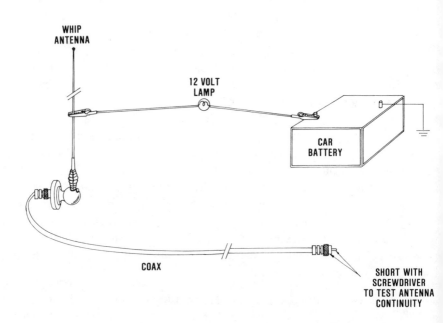

Figure 8-1: Test setup for a CB transmitter antenna open or shorted circuit test

Comments:

Sometimes it's very difficult to get to antenna feed lines. One of the easiest ways I've found to check an inacessible coax transmission line without instruments, is this simple procedure.

Procedure:

Step 1. Disconnect the coax cable going to the antenna from the CB transceiver by unplugging the cable's coaxial connector that goes into the antenna jack on the back of the tranceiver. *Note: Under no circumstances should you trim the coax without consulting the manufacturer's instructions.* In many systems, the transmission line is "tuned" for a specific length, to present a proper impedance.

Step 2. Connect one alligator clip to the antenna whip and the other to the car battery's "hot" terminal, as shown in Figure 8-1.

Step 3. Use a screwdriver or clip lead and short between the coax connector center pin and outside shell. If the lamp does not glow, the coax line, one of its connectors, or base connection is open.

Step 4. It's also possible that the lamp will light before you short the coax plug; in other words, just as soon as you connect the bulb, as shown in Figure 8-1. In this case, you may have a short in the antenna system. It could be the coax cable, antenna, or one of the connectors is shorted (another reason is explained in Step 5). I always check the coax connectors first, because they have proven to be more troublesome than the antenna or cable.

Step 5. The other reason your lamp may light is that the antenna is a shunt-fed, base-loaded type. This type system is easy to spot because it usually has a loading coil clearly visable in the whip. You'll have an operating lamp (with or without the coax plug shorted) with this system. See Chapter 11 for more details on testing antenna systems.

8.2   AM MODULATION MEASUREMENTS

QUICK MODULATION CHECK

Test Equipment:

Lamp-type dummy load (see section on dummy loads in this chapter)

Test Setup:

Connect a lamp-type dummy load to the transceiver antenna jack.

Procedure:

Step 1. Energize the transmitter and note the brilliance of the lamp as you speak into the mike.

Step 2. If the transceiver is modulating the RF carrier, you should see the lamp flicker. It should become brighter and dimmer, depending on the loudness of your voice (in AM sets). With a little bit of practice, you can learn to make an estimate of the percent of modulation.

VOLTMETER METHOD

Test Equipment:

RF voltmeter, dummy antenna, audio-signal generator

Test Setup:

See Figure 8-2.

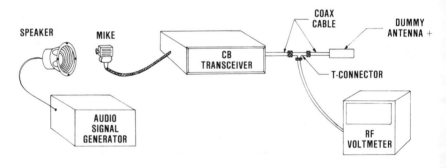

Figure 8-2: RF voltmeter connected to a CB transceiver and dummy load, through a T-connector

Comments:

In most cases, you'll find that the transmitter modulation is okay if the receiver audio is operating properly, because both use the same stages. However, if you do have trouble, using this method of checking modulation is one way to get around the problem of not having an oscilloscope. Furthermore, it's even possible that you may not need any additional equipment if you already have an RF ammeter in the system. In either case, with an RF voltmeter or RF ammeter, all you have to do is note the meter readings during modulation and refer to Table 8-1 to determine the percent of modulation.

ANTENNA CURRENT INCREASE WITH MODULATION	PERCENT OF MODULATION
22.5%	100%
18.5%	90%
15.5%	80%
11.5%	70%
8.6%	60%
6%	50%
4%	40%
2.2%	30%
1%	20%
0.25%	10%

Table 8-1: RF voltage (or current) increase VS percent of modulation

Procedure:

Step 1. Connect the equipment as shown in Figure 8-2.

Step 2. Modulate the transmitter by feeding 1,000 Hz acoustically into the mike (highest permitted modulation frequency is 3,000 Hz). Increase and decrease the output amplitude control on the signal generator.

Step 3. Observe the voltmeter (or ammeter) reading as you vary the signal generator output level control.

Step 4. Refer to Table 8-1 to determine the percentage of modulation.

Step 5. If the set overmodulates (more than a 22.5% voltage increase) on loud tones—a common trouble—check the AMC adjustment. *Note: Overmodulation may cause severe television interference during operation of the CB rig.*

Step 6. Weak modulation is an indication of mike or transistor troubles. Disconnect the mike cable and inject your 1,000 Hz test tone (set at about 35 mV rms for 100% modulation) into the mike input jack. If you can't get a 22.5% increase in your voltage (or current) reading between no-tone (zero voltage input) and a loud tone (about 35 mV rms input), it's a mike-amplifier or AMC defect.

OSCILLOSCOPE METHOD—SINE-WAVE PATTERN

Test Equipment:

General purpose oscilloscope, radio receiver that tunes to the CB frequency of interest, small RF coupling capacitor, and an audio-signal generator (optional).

Test Setup:

Connect the scope's vertical input terminal to the receiver's last IF amplifier transistor collector lead, using a DC blocking capacitor in series with the scope's "hot" lead, as shown in Figure 8-3.

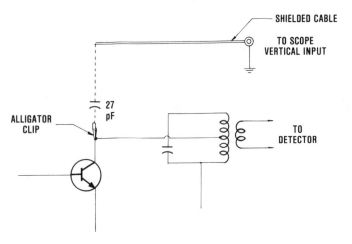

Figure 8-3: Scope connection using a radio receiver to check for overmodulation in a CB transmitter

Comments:

To actually "see" the RF carrier on an oscilloscope, requires a scope with a vertical input bandwidth up in the 30 or 40 MHz bandwidth region. Of course, this puts the scope in a price range out of reach of most of us. However, there is a way out of the dilemma. Use a down-converter. In other words, pass the modulated carrier through a receiver mixer and look at the lower intermediate frequency at the receiver's last IF amplifier. *Note:* It is all but impossible (except for a very rough estimate) to measure percent of modulation, using this procedure. However, what is most important, you can quickly detect *overmodulation*, which is what makes the FCC unhappy. Figure 8-3 illustrates some of the patterns you'll see on your scope. See Test 10.3 for another method of making a voltmeter measurement.

Procedure:

Step 1. Connect the oscilloscope vertical input test lead to the receiver's last IF amplifier collector lead, through an RF coupling capacitor.

Step 2. If you're testing the transceiver in-shop, place a dummy antenna on the antenna terminals *before energizing the set.* Don't worry about picking up the signal on the receiver being used for a down-converter because there should be ample radiation anywhere in your shop. If not, connect any piece of wire (for example, an unused test lead) to the receiver as an external antenna.

Step 3. Key the transmitter and apply sinusoidal modulation. Modulation can be done by acoustical means (through the mike), or with an audio-signal generator connected to the mike input of the transceiver under test. If you use a signal generator at the mike input, set the frequency for 1,000 Hz, with an output level of about 35 mV.

Step 4. Check the pattern on the scope. The most important ones you'll see are shown in Figure 8-4.

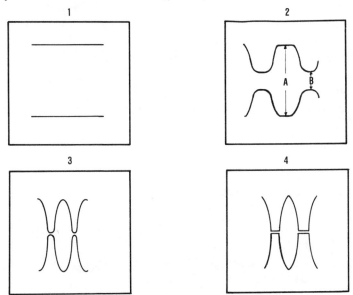

Figure 8-4: Approximate modulation patterns you'll see on an oscilloscope under various conditions. (1) RF carrier, no modulation, (2) less than 100% modulation, (3) 100 % modulation, (4) overmodulation

Step 5. A rough approximation of the percent of modulation can be calculated by using the formula:

$$\text{percent of modulation} = \frac{A - B}{A + B} \times 100$$

where A and B are as shown in Figure 8-4 (2). What you need is to be sure that the set is capable of more than 70% but not over 100%, the modulation is linear (appears equal on both top and bottom), and is sinusoidal.

8.3 DUMMY LOADS, CONSTRUCTION AND USE OF

LAMP-TYPE

Test Equipment:
Number 47 lamp and phono plug

Test Setup:
See Figure 8-5.

Comments:
Do not make internal adjustments on any CB transceiver when it is connected to a radiating antenna. Always use a dummy antenna. A simple one can be constructed using a pilot lamp and phono plug and may be used to terminate the CB rig during testing.

Procedure:
Step 1. Solder the circuit together, as shown in Figure 8-5.

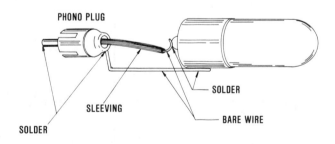

Figure 8-5: Construction details for a lamp-type dummy antenna

Step 2. Plug the completed dummy antenna into the transceiver antenna jack, for testing. The lamp acts as a load of approximately 50 ohms. It also will provide you with a means of checking relative power output and modulation (see Test 8.2).

COAXIAL TYPE

Test Equipment:

Coaxial plug PL-259 (sometimes called a *VHF plug*) and a ½ watt composition resistor

Comments:

The reason for the copper disk shown in Figure 8-6 is to reduce the lead inductance and provide shielding.

Construction Details:

Step 1. Select the correct value composition resistor for the make (or makes) CB set to be serviced (usually, approximately 50 ohms).

Step 2. Mount and solder the resistor in the coaxial plug as shown in the cut-away in Figure 8-6.

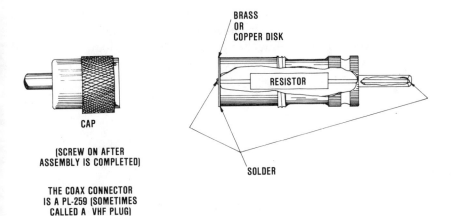

Figure 8-6: Dummy antenna for servicing CB transceivers, made using a composition resistor and coaxial plug. From "Practical Handbook of Low-Cost Electronics Test Equipment," by Robert C. Genn, Jr. © 1979, by Parker Publishing Co., Inc., West Nyack, New York

8.4 AUDIO LOAD RESISTOR POWER MEASUREMENT

Test Equipment:
Audio load resistor and rms reading voltmeter

Test Setup:
Substitute the audio load resistor for the speaker.

Comments:
In most cases, it is best to make receiver tests with the receiver terminated by a resistor, with resistance equal to the speaker impedance. The resistor should be rated at least 50% over the expected power dissipation.

Procedure:
Step 1. Refer to the service notes to find the rated power output of the equipment under test. This is the power that should be delivered to the audio load resistor.

Step 2. Measure the voltage across the load resistor with the rms reading voltmeter and use the indicated value to convert power to equivalent voltage, by using the formula $E = \sqrt{PR}$, where E is the rms voltage, P is the power listed in the service notes, and R is the load resistance in ohms.

Step 3. If you service quite a few two-way radio sets, it's well worth your time to use a calculator and work up a table, or graph, to convert power to rms voltage for the common audio loads (say 3.2 and 8 ohms). A sample table for an 8 ohm load resistor is shown in Table 8-2.

POWER WATTS	VOLTS (rms)
0.5	2
1	2.8
2	4
3	4.8
4	5.6
5	6.3

Table 8-2: With an 8 ohm resistor, this table can be used to convert required power readings to voltage readings

8.5 SIMPLE CB RADIO POWER
INPUT SYSTEM TEST

Test Equipment:
VOM

Test Setup:
Disconnect the CB set from the vehicle's electrical system or, if it's a base station operating off the AC power lines, pull the wall plug.

Procedure:
Step 1. Set your VOM to read on its lowest resistance range (for example, R x 1 or R x 10).

Step 2. With the set's ON-OFF switch turned at ON, measure the resistance across the CB unit power input leads.

Step 3. If your resistance reading is near zero, the CB transceiver power input system probably is operating correctly.

Step 4. If Step 3 checks out okay, your next step is to check the vehicle's electrical system (see Chapter 12). In the case of a base station, check for about 120 VAC at the wall plug. Incidentally, a quick way to check an automotive system is to connect the CB set directly to the car battery (see Comments, Test 5.16).

Step 5. On the other hand, if you read a very high resistance (or infinite resistance), in all probability there is a trouble in the power cable, plug, fuse, or the power circuit wiring inside the set.

8.6 MOBILE TRANSCEIVER SYSTEM BENCH TESTING

Test Equipment:
Power supply/battery eliminator, ammeter (at least 200 mA), voltmeter, and dummy antenna

Test Setup:
See Procedure.

Comments:
Inexpensive test equipment can be used for rough performance tests. However, the cables and controls used should duplicate the conditions of normal equipment use, or your measurements and tests will be further invalidated. For example, to simulate vehicular

installations, the transceiver should be connected to the ungrounded battery eliminator terminal with about 7 feet of No. 6 copper conductor. If you don't do this, the voltage drop on the vehicle cable contributes to the error and invalidates, to a greater degree, the measurement data obtained at the bench. This is most important when performing tests on trunk-mounted units.

Procedure:

Step 1. Connect the transceiver to the DC voltage work bench power supply. Watch polarity! Make sure the ground lead is not inadvertently connected to the "hot" power supply terminal.

Step 2. Place an RF dummy antenna on the transceiver antenna plug. You'll find construction details for a dummy antenna in this chapter (see Test 8.3).

Step 3. Place your ammeter between the bench power supply and transceiver power input cable. You can expect between 100 mA and 200 mA of current in a properly operating receiver. Therefore, set the VOM range switch to a range of 200 or above.

Step 4. Turn on the bench supply and set the output at 13.6V, ±0.5V.

Step 5. Note the receiver idling current (no signal being received) with the squelch open and the volume set at maximum. As was mentioned in Step 3, you'll probably measure a current between 100 and 200 mA, depending on the set. If your reading is above these values, it is possible that you have a shorted protective diode in the receiver. If you have to replace a diode, typically you'll find that they are rated at 1.5 amps with a peak inverse voltage (PIV) rating above 300 VDC.

8.7 EASY-TO-MAKE TRANSMITTER CARRIER FREQUENCY CHECKS

RADIO RECEIVER METHOD

Test Equipment:

Multiple-band calibrated radio receiver, DC voltmeter (the voltmeter is not needed if the receiver has an S-meter), and an RF dummy load (construction details for a dummy antenna are given in Test 8.3).

Test Setup:

Place a dummy antenna on the CB transmitter and loosely couple the transmitter to the receiver.

Comments:

If you don't have a suitably calibrated radio receiver with a built-in S-meter, it's possible to connect a DC voltmeter between the AGC bus line and ground. The voltmeter will act as a tuning indicator (S-meter). Or, you can contact a local amateur radio operator. It's very probable that he will help you, assuming that you only want to check your own CB rig. Many of the radio receivers used by amateur radio operators have a self-contained S-meter and crystal-controlled oscillator that provide checkpoints for receiver calibration. However, the simple frequency check described here does not meet the FCC technical standards, as defined for FCC record purposes. It is important that you realize that a transceiver *must remain within 0.005% tolerance* (approximately 1850 Hz), to comply with FCC rules and regulations.

When using an S-meter, be careful how you interpret the readings. For example, depending upon the rig, an S-meter may be calibrated so that one S unit is equal to 6 dB. Some S-meters, however, are calibrated at only 3 dB per S unit, and others at 3 or 4 at the low end and 6 or 7 at the top of the scale. Just remember, if you use different S-meters, your readings may vary considerably.

Another problem encountered with the S-meters, is the inability to measure high strength inputs. When checking a powerful signal, some S-meters will appear erratic in operation and even read lower on the scale, as the signal level is increased.

Procedure:

Step 1. Place a suitable dummy antenna (usually 50 ohms) on the CB transceiver under test. Use a nominal supply voltage (13.6 VDC for mobiles, 117 VAC for base stations), and conduct the test at room temperature (70° F).

Step 2. Energize the receiver and CB transceiver and let them warm up for 30 minutes or so. It isn't necessary to energize the transmitter during the warm-up.

Step 3. If the receiver has a self-contained calibrator (for best results, it should have one), calibrate the receiver.

Step 4. Energize the CB transmitter and tune the receiver for peak deflection on the S-meter, using loose coupling between the two (as much distance between the two pieces of equipment as practical).

Step 5. Read the CB transmitter frequency directly from the receiver tuning dial.

FIELD-STRENGTH METER METHOD

Test Equipment:
Frequency calibrated field-strength meter and dummy antenna

Test Setup:
Place a suitable dummy antenna (usually 50 ohms) on the CB transceiver and loosely couple the transceiver to the field-strength meter. Use a nominal supply voltage (13.6 VDC for mobile, 117 VAC for base stations), and conduct the test at room temperature (70° F).

Procedure:
Step 1. Set the field-strength meter and transmitter to the frequency to be measured.

Step 2. Turn the transmitter on and adjust the field-strength meter for peak deflection. If you're using antenna coupling (some field-strength meters use more than one method of coupling; i.e., capacitance, cable, etc.), use loose coupling. This can be achieved by varying the distance between the equipment.

Step 3. When you have the field-strength meter tuned exactly to the transmitter frequency (indicated by peak deflection), read the frequency off the field-strength meter tuning dial.

8.8   TRANSMITTER POWER OUTPUT MEASUREMENT

Measurement Equipment:
RF wattmeter (see Comments)

Measurement Setup:
See Figure 8-7.

Comments:
When servicing CB transmitters, it is better to use one of the bi-directional power meters that are designed to meet the needs of a CB

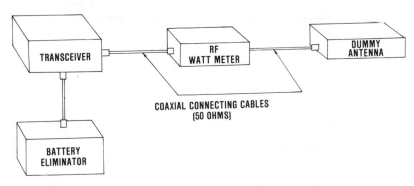

Figure 8-7: Test setup for measuring a CB transmitter's output power

serviceman. For example, they are capable of measuring low forward power (20 mW to 30 mW), and will also measure the reflected power, whereas many power meters won't budge off zero scale reading at 4 watts forward power.

You'll also find that you may need three additional pieces of test gear, depending on the power meter and test setup you have to use: (1) a dummy antenna (usually, 50 ohms), (2) a battery eliminator or other type DC power supply if you're testing on the bench, and, (3) an N-type coax connector to PL-259 adaptor in your coax cable connections.

Procedure:

Step 1. Connect the equipment, as shown in Figure 8-7. If the power meter has an internal dummy load, you will not need the external one shown.

Step 2. Measure the power output with the RF power meter (if you're servicing single-sideband (SSB) equipment, skip Steps 3 and 4).

Step 3. If the output power is below 3 watts (unmodulated), the transmitter is in need of repair or alignment. Try peaking the tuned circuits in the driver and final stages.

Step 4. If the output is over 4 watts (unmodulated), *under no circumstances* should the transmitter be connected to a radiating antenna because this is a direct violation of FCC regulations. In this case, measure the DC supply voltage. You should not have more than 14 VDC. If it's higher, adjust it back to 13.6 volts.

Step 5. When checking a CB transmitter (SSB), you should use a peak-reading RF wattmeter and measure "peak envelope power" (PEP). It is important to remember that PEP output is modulated power and, therefore, much greater than unmodulated power. However, what you want to read is 12 watts PEP on SSB. Anything over this is an FCC violation.

8.9 CRYSTAL TEST

Test Equipment:
Two operating CB sets

Test Setup:
See Procedure.

Comments:
Many inexpensive CB sets are designed to use a frequency synthesizer; the most popular being a 6-4-4 type. Figure 8-8 shows a

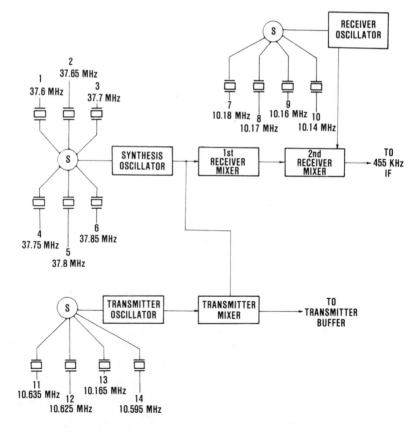

Figure 8-8: Block diagram of the most frequently used synthesizer in low-cost CB units

block diagram of a typical synthesizer, listing the crystal frequencies, and Table 8-3 lists the crystal combinations.

CHAN-NEL	CRYSTAL		CHAN-NEL	CRYSTAL		
1	7	11	13	4	7	11
2	8	12	14	4	8	12
3	9	13	15	4	9	13
4	10	14	16	4	10	14
5	7	11	17	5	7	11
6	8	12	18	5	8	12
7	9	13	19	5	9	13
8	10	14	20	5	10	14
9	7	11	21	6	7	11
10	8	12	22	6	8	12
11	9	13	23	6	10	14
12	10	14				

Table 8-3: Crystal combinations used in a 6-4-4 synthesizer

Procedure:

*Step 1.*Check the performance of two CB rigs. This can be a base station and mobile, two hand-held sets, or any combination of two similar CB transceivers.

*Step 2.*After you're sure that the two sets are operating properly, simply remove one of the internal crystals and replace it with the crystal of unknown quality.

Step 3. Again, check and see if you can establish communications between the two. If you can, the crystal may be okay. To make a positive test, measure the transmitter, or receiver, performance and, *above all, measure the set's frequency before placing it on the air.* Failure to do this, can result in an FCC citation for off-frequency operation. Be sure and measure the RF output frequency on every channel (the easiest way to do this is to use a frequency counter). All should be within 1 kHz of the assigned carrier frequency.

8.10 RECEIVER CRYSTAL CHECK

Test Equipment:
Properly operating CB transceiver, frequency calibrated variable frequency RF oscillator

Test Setup:
See Figure 8-9.

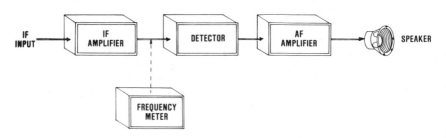

Figure 8-9: Test setup for checking receiver crystals

Comments:
Typically, a so-called *low-cost precision RF signal generator's* calibration accuracy is ± 3%, which is not good enough to permit use of one of these signal generators as a frequency standard. Although it probably is now listed as a museum piece, one signal generator (frequency meter) that can be used is a BC-221. The best place to find one of these is to write the dealers in surplus electronics, listed in the back of trade magazines such as *Popular Electronics, Radio Electronics,* etc. Or, you may be able to pick one up at a bargain price at one of your local surplus dealers. In either case, they are handy to have in the shop and will do the job (assuming they are properly calibrated).

Procedure:
Step 1. Open the CB reciever under test and loosely couple a frequency meter (such as the BC-221 mentioned) to the last IF, as shown in Figure 8-9.

Step 2. Energize the signal generator, CB receiver under test, and a separate CB transmitter, all tuned to the same frequency.

Step 3. Tune the signal generator until you have zero beat.

Step 4. Note the frequency reading on the frequency generator and compare it to the receiver manufacturer's published IF frequency (usually 455 kHz or 1650 kHz).

Step 5. In 6-4-4 synthesizers, four crystals commonly are used in the receiver oscillator. If the measured IF differs widely from channel to channel, it's an indication that the receiver crystals may not be correctly matched. A 6-4-2 unit (not as popular as a 6-4-4) uses one transmit crystal and one receive crystal. Channels that are on frequency for the transmit mode are accurate for receive, if the receiver oscillator is operating properly.

8.11   RECEIVER AUDIO STAGE TEST

Test Equipment:
Audio-signal generator and a good quality AC voltmeter

Test Setup:
See Procedure.

Comments:
In most cases, you'll find the receiver audio stages are working properly, if the transmitter modulation is found to be correct, because both use the same stages.

Procedure:
Step 1. Set the audio-signal generator to operate at 1,000 Hz and connect it to the output side of the AM detector diode.

Step 2. With the receiver under test, and the signal generator turned on (and connected as explained in Step 1), measure the output of the signal generator and set it to about 0.2 volts rms. You can use a scope to make this setting (about 0.5 volts peak-to-peak).

Step 3. Measure the rms voltage across the speaker terminals.

Step 4. Next, use the formula $P = E^2/R$ to calculate the power delivered to the speaker. You'll probably have slightly more than 3 watts. For example, with an 8 ohm speaker and a 5 volt rms voltmeter reading:

$$P = E^2/R = 5^2/8 = 3.125 \text{ watts}$$

8.12   SQUELCH TEST

Test Equipment:
RF signal generator that can produce a modulated RF signal level of about 1 μV

Test Setup:
See Procedure.

Procedure:
Step 1. With no modulated RF signal, set the squelch control so that it just cuts off all noise.

Step 2. Couple the RF signal generator to the transceiver antenna and apply a modulated signal of about 1 μV. If the squelch is operating properly, the modulated signal should cause the receiver to again start to operate.

Step 3. The squelch control knob should be set at about 1/3 rotation, to cut off audio with no signal applied. If it is not, loosen the knob set screw and set it to this position.

8.13   AUTOMATIC GAIN CONTROL (AGC) TEST

Test Equipment:
RF signal generator that can produce a modulated RF signal from 5 μV to 100,000 μV (for example, Heathkit model IG-5242, or any similar one), and an AC voltmeter or scope

Test Setup:
See Procedure.

Procedure:
Step 1. Set the signal generator to produce a modulated RF output signal in the channel the receiver under test is tuned to, and adjust its output level for 10 μV.

Step 2. Couple the generator's output signal to the CB transceiver antenna.

Step 3. Connect an AC voltmeter across the speaker terminals and measure the audio voltage (don't disconnect your voltmeter).

Step 4. Increase the signal generator RF output level to 10,000 μV.

Step 5. Again, note the voltmeter reading. Now, the readings in Step 3 and Step 5 (this Step) should not show an increase of more than 10%, if the AGC is operating properly.

8.14   PHASE-LOCKED LOOP (PLL) TEST

Test Equipment:
VOM and good quality (well-filtered, well-regulated) DC power supply and a frequency counter

Test Setup:
See Procedure.

Comments:
The following procedure is designed to test most typical, low-cost, 23 channel CB transceivers that incorporate a PLL in their design.

Procedure:
Step 1. Measure the DC voltage on the input of the voltage controlled oscillator (VCO). Refer to the set's schematic for the correct voltage. Typically, the voltage will be in the 1 to 2 volt range. If this voltage is off more than 4 or 5 tenths of a volt, it can cause the VCO to operate off frequency and, in severe cases, kill the oscillator completely.

Step 2. Connect a DC power supply to the VCO DC control voltage line (between the phase detector and VCO). Also, connect your VOM to the same point (many sets have a test post in this line).

Step 3. Using the VOM, set the power supply output voltage to the level indicated on the set's schematic. Again, it will probably be between 1 and 2 volts.

Step 4. Next, connect a frequency counter to the VCO output.

Step 5. Refer to the schematic. Your counter should read within the limits set by the manufacturer. You'll find these PLL's should operate near several different center frequencies. For example, sets using PLL-01AIC's operate the VCO at a frequency of 21 MHz. Sets using a separate reference oscillator, operate at 6.4 MHz, and the PLL-02A (and most others) operate at about 38 MHz, using a 10.24 MHz frequency reference.

If you can't get the VCO to operate at the proper frequency by adjusting the power supply to the voltage shown on the schematic, it's an indication that the VCO needs adjustment or replacement.

Step 6. Next, check the loop stage-by-stage, referring to the schematic and/or service manual. However, servicing is made much easier if you can get the system to operate at any frequency, using the DC power supply and adjusting its output to whatever voltage is needed to start the system operating. Before you start troubleshooting, try a voltage slightly higher and slightly lower. It may start the system operating. Once you get it going, all that's left to do is refer to the schematic for the correct voltages to locate the stage at fault. A block diagram of a typical PLL is shown in Figure 8-10.

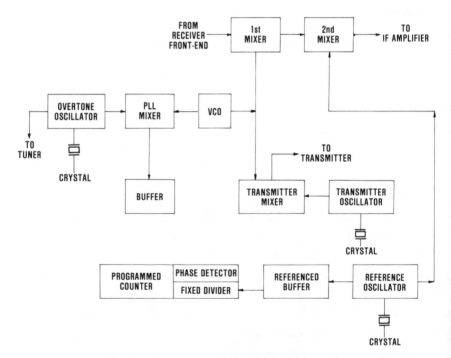

Figure 8-10: Block diagram of a typical 23 channel PLL system

Chapter 9

Electronics Tests and Measurements
For Remote Control Equipment

This chapter offers a full collection of tests and measurements that can save you hours of time and effort when working with remote control systems. Furthermore, the following tests and measurements can give you the extra edge that's so important in today's constantly changing electronics field. In fact, in no time at all, you'll be able to test numerous control devices, because this chapter gives you the information you need. . .step-by-step. It will increase your effectiveness in the work shop, thus giving you a more profitable electronics operation.

Field-strength testers are expensive, right? Not necessarily. Why not build your own? It's easy. A little bit of hook-up wire, a pencil and a neon bulb are all you'll need. Furthermore, this simple field-strength checker can be used to check the RF output of radio transmitters, garage-door-opener transmitters, RF type TV remote control transmitter, low-power CB walkie-talkies, and model aircraft RF control systems. You'll find many practical suggestions, both at the transmitting and receiving end, discussed in the following pages.

9.1 TV REMOTE CONTROL PRELIMINARY TEST

Test Equipment:
   None

Test Setup:
   None

Comments:

Remote control transmitters used with TV receivers normally have buttons to press that turn the set on and off, switch channels, show channel identification, and control volume. Typically, there are between 8 to 15 buttons, depending on how many functions are controlled.

Almost all systems use a frequency in the ultrasonic region, normally between 35 kHz and 45 kHz. The transmitter radiates a different frequency for each function. For example, an 8-function control unit must produce eight different frequencies (8 channels). One unit (produced by Magnavox) generates 15 control signals. In this digital system, the receiving unit in the TV receiver counts the incoming frequency to decode and identify the function, and then the logic section determines what function the signal controls.

Procedure:

Step 1. Press all function buttons on the remote control to determine if any of the controls will activate any of the systems in the TV receiver.

Step 2. If any of the control keys produce the labeled result, it means that the transmitter battery(if there is one), both transmitter and receiver transducers, transmitter oscillator and receiver oscillator (if there is one), are all functioning properly. To troubleshoot the system, merely check the non-operating channels. However, during your troubleshooting, don't overlook the possibility that the TV receiver may need alignment.

Step 3. If you can't get any one of the responses, your next step is to determine if the problem is in the transmitter or receiver. The easiest way to do this is to substitute a known-to-be-good transmitter. If you don't have a replacement transmitter, your next best step is to refer to the manufacturer's schematic for the particular chassis you're working on. Fortunately, a lot of digital systems are divided into separate logic blocks or functions. You should be able to pin down the function or stage that is not working properly and replace it by changing an integrated circuit, or by replacing a module. For example, in a TV tuner with a digital readout and varactor tuning, if the set tunes to the proper station but the readout doesn't work, or shows the wrong channel, you've narrowed it down to a problem in the display section.

As another example, if the "tune lower" or "tune higher" function of a TV tuner isn't working, you can use the same system you use to check a stereo. Check the channel that isn't working against the one that is. Make a cross-check to see what the DC levels should be at each point. Figure 9-1 shows a block diagram of an LSI chip (used by Magnavox in one of their remote control receiver units) that you can check using these suggestions.

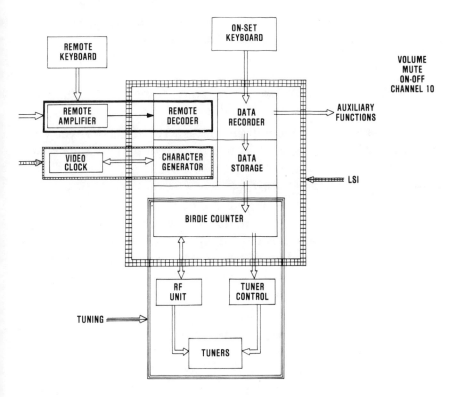

Figure 9-1: Block diagram of an LSI chip in a TV remote control receiver. See text for a method to check a system using an LSI such as this. From the book, "Workbench Guide to Electronic Troubleshooting" by Robert C. Genn, Jr., © 1977 by Parker Publishing Co., Inc., West Nyack, New York

9.2 BATTERY TEST

Test Equipment:
 VOM

Test Setup:
  See Procedure.

Comments:

There are remote-control transmitters that have self-checking systems which indicate battery chassis condition; for example, the Admiral ERA 11 chassis. To check the battery, simply press the *select* button and, if the battery is strong, you should see the least significant digit on the receiver's digital display begin to flash. When the battery is weak, the lower digit stops flashing. On remote control units without a self-checking system, use the following procedure.

Procedure:

Step 1. Set your VOM function switch to read a DC voltage and the range switch to the 10 VDC position.

Step 2. Connect the VOM test leads across the battery you want to check. It's best to leave the battery in the transmitter when making this test.

Step 3. Press each button on the transmitter keyboard and observe your voltmeter reading. If the battery is strong, you should read between 7 and 8 volts on all function switches, in most cases. In general, if you read a full battery voltage, it's an indication there is an open circuit, and a reading of zero voltage indicates a short circuit (assuming the battery is good). If you see either of these readings as you press the buttons, the transmitter is defective.

9.3 TRANSDUCER TESTS

VOLTMETER METHOD

Test Equipment:

VTVM or equivalent voltmeter

Test Setup:

Connect the voltmeter test lead across the transducer, as shown in Figure 9-2.

Comments:

Some remote control systems use exactly the same transducer in the transmitter and receiver. If this is the case, a good check can be made by interchanging one with the other.

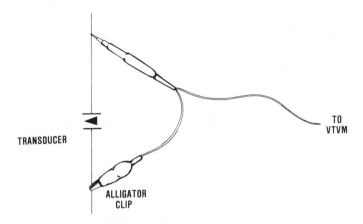

Figure 9-2: Test probe connection for measuring the ultrasonic signal voltage on a TV remote control transducer

Procedure:

Step 1. Connect a voltmeter across the transducer, as shown in Figure 9-2, and set the voltmeter function switch to read AC. Set the voltmeter range switch to read 1,000 VAC full-scale. You'll probably have to drop down to a lower scale. However, it's best to start high, to be on the safe side.

Step 2. Press each function button on the remote control. You should read a large voltage (a TV remote control will be somewhere in the hundreds) as you press each button, if you're reading a peak-to-peak scale, or about 110 volts or so, rms. Refer to the manufacturer's service notes, or another properly operating transmitter, for the voltage that should be measured on the particular unit you're servicing (300 volts peak-to-peak is typical in TV systems). If you measure a very low voltage (or no voltage), it indicates the associated circuit is not working, or it's possible that a series capacitor is at fault. See Test 9-4.

OSCILLOSCOPE METHOD

Test Equipment:
Oscilloscope

Test Setup:
None

Procedure:

Step 1. Set your scope controls to measure an ultrasonic frequency somewhere between 35 kHz and 45 kHz. The actual frequency will depend on which button you press on the remote control unit. Turn the scope vertical gain up to maximum.

Step 2. Hold the remote control unit close to the scope's vertical input terminal. Press each control button on the unit and you should see a signal on the scope for each function. Remember, each function uses a different frequency, so you'll probably have to re-adjust the scope each time if you want to examine the waveform.

9.4 TRANSDUCER CAPACITOR CHECK

Test Equipment:
Voltmeter (VTVM or equivalent)

Test Setup:
Connect the AC voltmeter across the transducer's series capacitor.

Comments:
The typical color TV remote control units of older color sets (early 1970's) use a capacitor in series with the transducer. If a voltage check across the transducer reads zero, the capacitor should be checked by using the following procedure.

Procedure:

Step 1. Connect an AC voltmeter across the series capacitor. Set the voltmeter to read 1,000 volts to start, then bring it down if necessary.

Step 2. Press each function switch on the remote control unit. Typically, you'll read about 106 volts rms or 300 volts peak-to-peak across the capacitor, if it is good; i.e., not leaky or shorted.

Step 3. If your reading seems low, try paralleling a new capacitor across the one in the circuit. Again, press the button and monitor the voltmeter. If you read a higher voltage, or if you can get the unit to activate a TV receiver, replace the capacitor.

Step 4. If you can't get results by paralleling the capacitor, check the preceding transistor.

9.5 REMOTE CONTROL TRANSMITTER
ALIGNMENT TEST

Test Equipment:

Remote control unit that is known to be good (of the same type as the one under test) or an audio-signal generator that tunes over the frequency range of the unit (usually 35 kHz to 45 kHz), and an oscilloscope

Test Setup:

See Procedure.

Procedure:

Step 1. Connect the signal generator to the scope's vertical input. Or place your reference control unit near the scope's vertical input. Set the scope's vertical gain to maximum.

Step 2. Press the same two buttons on both the remote control units while holding them near the scope's vertical input. When using a signal generator, adjust it to zero beat with the unit under test (see next Step).

Step 3. If the two signal sources are transmitting the same frequency, you should see the two signals produce a zero beat on the face of the scope. Do this for each function and adjust the trimmer capacitors in each channel, if you can't get a zero beat. Figure 9-3 shows the zero beat patterns you should see on the scope when both

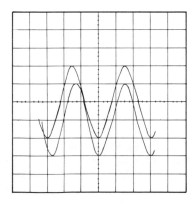

Figure 9-3: Remote control transmitter frequency check scope patterns at zero beat

frequencies are the same and the scope is set up correctly. As a general rule, you'll find it's better to start at the highest frequency (about 45 kHz) and then do the lower frequencies.

9.6 TV REMOTE CONTROL RECEIVER
ALIGNMENT TEST

Test Equipment:
    Remote control transmitter (or audio frequency generator) and VTVM or equivalent

Test Setup:
    See Figure 9-4.

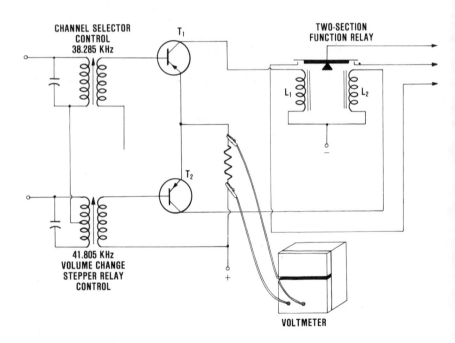

Figure 9-4: An example test setup for testing relay control circuit alignment in a TV remote control receiver

Comments:
    Receiver alignment problems can show up in a variety of ways. For example, you may be able to make the system work on channel selector or on volume control, but not on both at the same time.

Generally speaking, the best way to handle problems like this is to consult the alignment procedure published by the manufacturer. However, receiver alignment testing is quite easy even without service notes.

Procedure:

Step 1. Connect a VTVM, set to read a low DC voltage, across the emitter resistor, as shown in Figure 9-4.

Step 2. In many sets, you'll find two tuned transformers and a 38 kHz transformer comprise the channel selector. Check the alignment of each of these transformers and make sure the adjustment shows a definite peak reading on your voltmeter.

Step 3. If your alignment check seems to be correct, it may be that one of the ringer bars in the transmitter is broken, or something is damping it and throwing it off frequency.

9.7   RINGER BAR TRANSMITTER TEST

Test Equipment:
  None

Test Setup:
  See Procedure.

Comments:
  Several manufacturers use ringer bars that produce supersonic tones (for example 38 kHz to 45 kHz) when struck by a hammer in the transmitter unit. The system works something like a piano where hammers hit the steel wires, except that you press the buttons on the remote control transmitter rather than piano keys.

Procedure:

Step 1. Press the activating buttons on the remote control unit. You should hear a tone if the hammer for each function strikes the ringer bar and the bar is operating. The tone you hear *is not* the supersonic tone that activates the receiver functions, but it is an indication that the bar is at least operating in the audible frequency range.

Step 2. If you hear a tone and the system still does not work on one or more positions, it's possible that something is damping one of the ringer bars, or the receiver needs alignment. See Test 9.6.

9.8   TV REMOTE CONTROL MOTOR TEST

Test Equipment:
VOM

Test Setup:
See Procedure.

Comments:
There are several problems that might appear as motor troubles. For example, the motor may run past the wanted station, or stop at unwanted stations. If the motor does either of these, of course, the motor is operating properly, therefore it must be a circuit trouble. On the other hand, the motor can be stopped cold, and it still could be a circuit trouble. Again, it is necessary to check the transistors and other components.

Procedure:
Step 1. Check the motor bearings for looseness, binding, excessive dirt, and proper lubrication.

Step 2. Set the VOM to measure 115 VAC and check the AC line voltage on the input to the motor (you'll find 24 volts on the motor leads in some receivers). If you measure the correct voltage on the motor leads, but the motor won't operate, perform Steps 3 and 4.

Step 3. Remove all power and disconnect the motor leads.

Step 4. Set the VOM to read resistance and measure the resistance of the motor windings. If the winding is burned open, your ohmmeter will read infinite resistance. If the windings are partially shorted, however, you'll have a resistance reading. In this case, it's better to use a scope and make a ringing test, or you can use a low-ohmmeter. If you use an ohmmeter, you have to know the correct value of resistance when you suspect partially shorted turns. To find this value, either compare your reading to a known-to-be-good motor of the same type or consult the service notes.

9.9   PHASE SHIFT SYNCHRONOUS MOTOR TEST

Test Equipment:
Jumper lead

Test Setup:
    See Procedure

Procedure:
    Step 1. Try different functions and if the phase shift synchronous motor turns either way, it's an indication that the motor and all the circuits for that function are good.

    Step 2. Place a jumper lead from an operating driver transistor's collector lead to the inoperative transistor collector lead. If the motor starts to turn, your trouble is the transistor or its associated components. Simply inject an input signal to the transistor input circuit to check this out.

9.10 MODEL AIRCRAFT DIGITAL-PROPORTIONAL TRANSMITTER TEST

Test Equipment:
    Oscilloscope and RF probe

Test Setup:
    See Procedure.

Comments:
    See Testing Garage Door Openers (Test 9.11), Transmitter Test, for a simple field-strength meter that can be used to check the RF output of small remote control transmitters.

Procedure:
    Figure 9-5 shows a block diagram of a typical, contemporary 5-channel, digital-proportional transmitter. Touch your scope probe to the points indicated and, if any stage does not show a pattern similar to the ones in the illustration, it's an indication that particular stage is not operating properly.

9.11 TESTING GARAGE DOOR OPENERS

TRANSMITTER TEST

Test Equipment:
    Field-strength meter (see Comments)

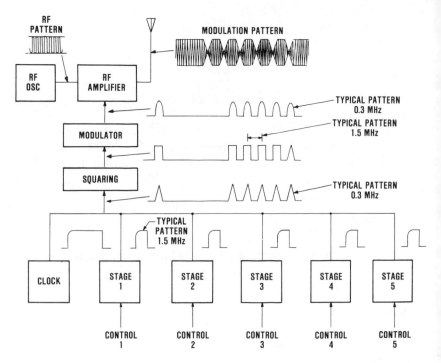

Figure 9-5: Oscilloscope patterns that should be seen when checking a digital-proportional transmitter

Test Setup:
        None

Comments:
        These transmitters usually are solid state and often crystal controlled. Example frequencies are 465 MHz and 41 kHz, and their transmissions frequently are coded. Digital circuits operate on a personal command frequency you set up yourself. You choose from many possible codes. With the switches set in the transmitter and receiver, the opener will operate on that frequency. You may change codes as often as you like (see the operating notes that come with the unit or contact a retail store that carries the particular brand, for code-setting instructions).
        To make a simple check to see if the transmitter is radiating an RF signal (this check does not insure that the code is correct), place a field-strength meter close to the transmitter. The field-strength meter can be nothing more than one made by the old trick of connecting a neon bulb to wire wound around a pencil, as shown in Figure 9-6.

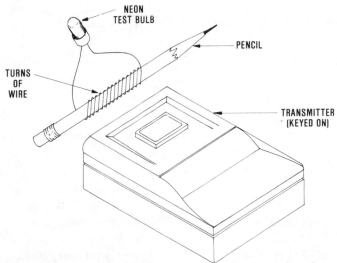

Figure 9-6: A simple field-strength checker that can be used to test for RF output on a garage door opener transmitter (or any similar device)

Procedure:

Step 1. Place your field-strength checker as close as possible to the transmitter antenna. This is the end of the transmitter you point toward the garage door during normal operation.

Step 2. Key the transmitter. In some cases, you may have to press the unlock tab on the transmitter before the receiver will respond to the operate command. Also, it's possible that you won't have the desired results (an opening door) because the emergency pull cord release has been pulled, which places the garage door in manual operation. Assuming that none of these possibilities exist, if the neon light glows you have RF being radiated by the transmitter.

Step 3. Perform the following Receiver Test because the transmitter probably is good. I say "probably" because the code could be in error. Just to be sure, double check your settings.

RECEIVER TEST

Test Equipment:

    None

Test Setup:

    None

Procedure:

Step 1. Check the transmitter for RF radiation, as explained in the preceding Transmitter Test.

Step 2. Try to operate the garage door opener by pressing the operations button on the unit mounted inside the garage (usually mounted in a convenient place on the wall).

Step 3. If the transmitter checks out okay and you can operate the garage door using the inside control, it is an indication that you have receiver troubles. Troubleshoot the receiver just as you would any solid state radio receiver. There is very little difference. You'll notice the receiver should stay operating in a standby mode at all times when not in use. This usually is done by placing a cut-off bias in the solid state components. If you can get the transmitter signal to overcome the receiver's cutoff bias when it is close to the receiver, but not at a distance, it's very probable that the transmitter RF output is weak, or you have a faulty component in the receiver. In this case, it's best to refer to the service notes. If you can't get the service instructions, simply troubleshoot the units like any small radio system; peak up the RF transformers, check DC voltages, etc.

MOTOR TEST

Test Equipment:

VOM

Test Setup:

See Comments, Procedure, and Figure 9-7.

Comments:

If a garage door will not open, your first step should be to use the emergency pull cord release that permits manual operation and check the door by hand. In some cases, you may have to remove the connecting arm from the door in order to permit manual operation. Other systems have a chain-drive opener that has a semi-automatic transmitter; i.e., you must keep constant pressure on the button to operate the opener. Whatever the system, it's best to first check the door by hand, if practical. Incidentally, some systems (such as Montgomery Ward's) have an unlock tab on the transmitter. If it isn't pressed, the receiver will not respond to the operate command.

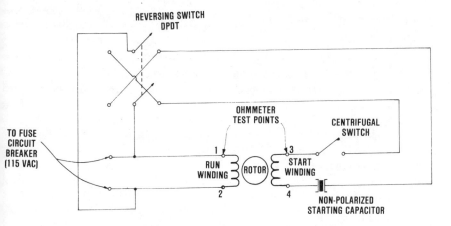

Figure 9-7: Wiring diagram and test points for a capacitor-start motor and its associated components

Procedure:

Step 1. Refer to Figure 9-7 when making the following tests. This illustration shows a typical capacitor-start motor and its associated components.

Step 2. Remove all power to the unit and connect your ohmmeter across the fuse or circuit breaker. If you read zero resistance, the circuit is good; infinite resistance, the fuse is blown or the circuit breaker is open.

Step 3. Use your ohmmeter and check the starting capacitor (see Chapter 4).

Step 4. Using the ohmmeter, check the run winding and start winding (test points 1, 2, and 3, 4). If a winding is burned out, you'll read infinite resistance. On the other hand, if the winding is partially (or completely) shorted, you'll see a low reading on the ohmmeter. In this case, you can make a low-ohmmeter measurement, as explained in Test 9.8 (TV Remote Control Motor Test), or make the ringing test mentioned in the same section.

Step 5. To check the double-pole, double-throw (DPDT) reversing switch, merely watch it during operation or see the following Relay Test for a detailed check. But if you have to service the electronics circuits, take care when changing RF bipolar transistors in some systems. For example, in one system (Genie garage door opener), you'll find the transistor is mounted on the PC board with the base lead pulled through the board between the collector and emitter lead. The important thing to note is that this mounting reverses the leads. Furthermore, you'll even find the transistor may show a little gain if connected the wrong way, which only adds to the confusion. In other words, check the transistor leads very carefully before pulling them.

RELAY TEST

Test Equipment:
    None

Test Setup:
    Remove the equipment case that contains the receiver relay and motor relay.

Comments:
    See Figure 9-7 for a typical relay wiring diagram.

Procedure:
    Step 1. Using the eraser end of a pencil, press the receiver relay points until they are closed. This should cause the motor relay contacts to close, and the motor to start operating.

    Step 2. Next, press the motor relay until its points make contact. This should cause the motor to start operating.

    Step 3. If the relay contacts are pitted or corroded, clean them (in severe cases, replace them). When checking relays, you only need an ohmmeter to determine whether the solenoid is open or shorted. Simply measure the resistance of the winding. If you read an infinite resistance, the winding is open; zero, or near zero, indicates the coil is probably okay, but be careful because it could be partially shorted. If you suspect it's partially shorted, see Test 9.8 (TV Remote Control Motor Test).

Chapter 10

Practical Guide to Medium Power
Transmitter Tests and Measurements

We are all interested in just how well modulated our transmitter is, and it's common knowledge that the closer a transmitter is to having 100% modulation, the better the received signal will be. But overmodulation can cause interfering signals (splatter) and, possibly, bring in an FCC "ticket" as well. This chapter includes a simple scope circuit that you can build which will permit you to skillfully monitor the modulation level of any AM transmitter—old or new—even if you don't own an oscilloscope.

The most important transmitter checks are power out, modulation, and frequency. You'll find all these explained in the following pages. This chapter will show you how to expertly test a radio transmitter—AM, FM, or SSB—using only simple and fast steps. You'll also find how to measure the performance of your transmitter, how to get the most out of it with the highest quality of performance and, in most cases, with inexpensive test equipment.

Included are easy-to-build coupling devices that can be used with all kinds of test gear, with the construction details clearly shown. There are oscilloscope patterns for checking SSB, AM and CW transmitters. Furthermore, in every test and measurement, you get complete step-by-step instructions that can save hours of time and effort when testing all kinds of transmitters.

10.1 SIMPLIFIED MEASUREMENT OF
UNMODULATED AM RF POWER OUT

Test Equipment:
Voltmeter with RF probe and a suitable dummy antenna (see Comments)

Test Setup:
    See Procedure.

Comments:
    Quite a few of the power measuring instruments on today's market are *directional* wattmeters; for example, the power/SWR tester sold by Radio Shack and the similar wattmeter/SWR bridge sold by Heathkit. While it's true that directional wattmeters are the best way to check antenna system performance, they aren't necessary for conducting standard medium power transmitter power measurements. All you really need is a voltmeter with an RF probe and a dummy antenna. Figure 10-1 shows a schematic of a commercially built, 50 ohm termination, that is readily available.

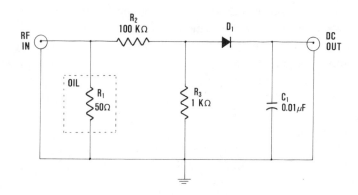

Figure 10-1: Schematic of Heathkit's dummy antenna. It handles up to 1 kW
with VSWR's of less than 1.5:1.

If you don't have an RF probe, it's easy to construct one. Probably, the simplest and best known way to build an RF probe is shown in Figure 10-2. The only thing that you'll have to be careful of is the voltage rating of the diode. The one shown is rated for a maximum peak reverse voltage of 75 V and a maximum forward voltage of 50 V, at a maximum forward current of 5 mA. To determine actual RF power out of the transmitter, you'll have to measure the voltage across the RF input to a dummy load. The following procedure explains how to do this.
    When using the probe and DC voltmeter, you'll find that accuracy (about 10%, under ideal conditions) is best near the top end of the scale. To help improve your accuracy, use as few connectors as practical in your connecting cables, and all cables should be assembled

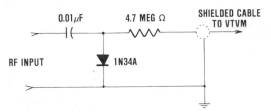

Figure 10-2: Schematic of an RF probe that can be used with a DC voltmeter to measure the peak value of an RF signal.

as carefully as possible. Be sure to remember when working with radio frequencies that every transmission line connector is a potential source of impedance discontinuity which can cause an increased SWR and power loss.

Procedure:

Step 1. Terminate the transmitter with a dummy antenna. Whatever you use as a dummy antenna, remember, this load must dissipate the power output of the transmitter under test for long periods, without physical damage or impedance changes.

Step 2. If the transmitter does not have a built-in, stable DC supply voltage, connect an external power supply to the unit.

Step 3. Connect the voltmeter's demodulation probe across the transmission line feeding the dummy antenna.

Step 4. Adjust (or re-check) transmitter tuning, loading, coupling and/or power adjust control settings.

Step 5. Read the voltmeter DC scale. Then use the formula:

$$\text{Power} = \text{voltage}^2 / 50 \text{ Ohms}$$

Warning: If you use the circuit shown in Figure 10-2 with your VTVM, it will read the *peak value* of voltage on the DC scale. To convert to rms voltage, use the formula:

$$\text{rms voltage} = 0.707 \times \text{peak voltage}$$

10.2 SIMPLE RF CURRENT MEASUREMENT

Test Equipment:

VOM, demodulator probe, 10 ohm carbon resistor (10 watts or more), and/or a dummy antenna (usually 50 ohms)

Test Setup:

See Procedure.

Comments:

If you don't have a resistor capable of handling the power output of the transmitter under test, simply use resistors in a series-parallel combination to achieve the required power rating and correct impedance.

Procedure:

Step 1. Place a 10 ohm, 10 watt carbon resistor in series with the ground side of the transmission line you're going to use to feed the dummy antenna. The reason for using the ground side of the transmission line is to prevent accidental shorts.

Step 2. Connect the transmission line and resistor network to the transmitter under test.

Step 3. Connect the demodulator probe across the resistor. Set the voltmeter to a low DC scale (about 10 volts midscale, in this case).

Step 4. Energize the system and measure the voltage, then convert your voltage reading to current, using the formula $I = E/R$. For example, let's say you read about 14 VDC. Now, this is peak voltage (using a peak-reading demodulator probe), so $0.707 \times 14 =$ approximately 10 volts. Therefore, $10/10 = 1$ ampere. Next, knowing the current, calculate the power.

$$P = I^2R = I^2 \times 10 = 10 \text{ watts}$$

To measure the higher power, use a higher power resistor and measure the voltage across the dummy antenna. As an instance, into a dummy load (50 ohms), with a voltage reading of 100 volts rms, the current is 2 amperes. The power comes out to be 200 watts.

10.3   AM MODULATION MONITOR AND MEASUREMENT PROCEDURE

Test Equipment:

"Home-brew" circuit shown in Figure 10-3

Test Setup:

See Procedure.

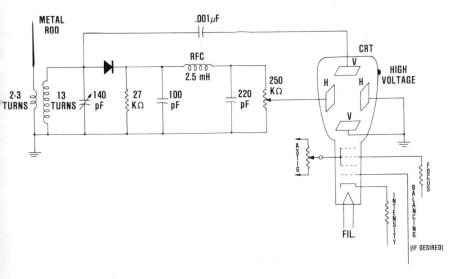

Figure 10-3: Circuit diagram for an AM modulation monitor that will produce a trapezoidal modulation pattern during use

Comments:

If you don't have an oscilloscope, you can build a suitable power supply for almost any electrostatic CRT (these usually can be found in electronic surplus outlets such as the ones in the back of *Radio Electronics,* etc.) that, when connected to the circuit shown in Figure 10-3, will function as a modulation monitor. Other systems of measuring modulation are described in Chapter 8, Section 8.2.

The coil (L_1) and capacitor (C_1) you use in the tuned input circuit are dependent on the frequency band the transmitter works in. The rest of the circuit is simply a diode detector and filter that couples the demodulated audio signal to the CRT horizontal deflection plates, plus one line and capacitor to the vertical plates. If the output of the transmitter is strong enough, you can get by without the antenna, its coil, and capacitor C_1. Almost any general purpose or high frequency diode (such as a 1N34A or 1N60) can be used in the circuit.

Procedure:

Step 1. Turn on the CRT power supply. You'll see a spot on the screen when it warms up. Incidentally, don't leave the spot on too long before going to the next step because it's possible that you might burn the CRT face plate.

Step 2. Key the transmitter and move the monitor antenna (or whatever you're using to couple the monitor to the transmitter) to several positions until the CRT spot changes to a vertical line. Adjust coupling until vertical line fills about two-thirds of the CRT screen.

Step 3. Speak into the microphone and you should see the line change into a trapezoid similar to one of the patterns shown in Figure 10-4. Use the 250 k-ohm pot to adjust the width of the pattern.

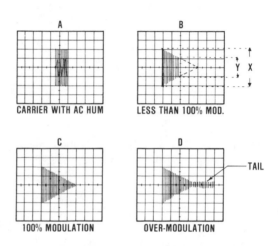

Figure 10-4: Trapezoidal patterns that may be seen while observing the modulation monitor described in the text

If you see the pattern shown in (A), you have AC hum problems (could be happening inside the transmitter or being picked up from the lights, etc., in the room), (B) is normal but showing less than 100% modulation (see the next step), and (C) is within legal limits and what you should see at 100% modulation. You want to be able to produce this pattern, or very close to it, for best reception at the receiver end. (D) is showing a "tail," which means the transmitter is being overmodulated and, if you radiate a signal under these conditions, it's sure to cause you much trouble with the FCC, your neighbors (it can cause television interference), and other radio operators, because of splatter.

Step 4. For this step, use Figure 10-4 (B) and the two points on the waveform marked (X) and (Y). To calculate the approximate percent of modulation, inject an audio signal into the transmitter microphone input and then measure the height of the two points. After you make the measurements, simply use the percent of modulation formula:

$$\text{percent of modulation} = \frac{X - Y}{X + Y} \times 100$$

As an example, let's assume you measure 2.5 inches for the larger vertical (X) and ½ inch on the smaller vertical (Y). Substituting the measurement values for the letters in the formula, we get,

$$\text{percent of modulation} = \frac{2.5 - 0.5}{2.5 + 0.5} \times 100 = 66\% \text{ modulation}$$

10.4 CW TRANSMITTER TELEGRAPH KEY MODULATION CHECK

Test Equipment:
Telegraph key, oscilloscope, and dummy load if needed

Test Setup:
Couple the scope to the AM transmitter output circuits. Place a dummy antenna on the system, if the transmitter is not connected to its normal antenna.

Procedure:
Step 1. After you've completed the test setup, turn on the scope and transmitter (don't key the transmitter), and let them warm up.

Step 2. Next, key the transmitter and make any necessary scope adjustments that are required to display the signal on the face of the CRT. The width of the patterns may be set to any convenient length. The height will depend on the RF power output of the transmitter and coupling. Don't worry too much at this point about the scope sweep frequency.

Step 3. Key the transmitter with a series of dots. The only way you'll get successful results (evenly spaced and stationary scope patterns) is by using an automatic key. While observing the scope, set the scope sweep frequency to a slow rate and adjust it until you see a stationary pattern similar to the one shown in Figure 10-5.

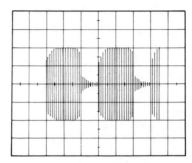

Figure 10-5: Oscilloscope pattern you should see from a plate modulated transmitter, using a keying filter and keyed with an electronic key

Step 4. Observe the scope patterns. They should be evenly spaced and not perfectly square (assuming there is a keying filter in the system). Ideally, each dot pattern should be rounded off at both the leading and trailing edges, as shown in Figure 10-5, not too much and not too little. Too much rounding off will cause the dots to run together and make it very difficult for the person at the receiving end to copy your message.

10.5 SSB POWER OUTPUT MEASUREMENT

Test Equipment:

Oscilloscope, dummy antenna, and audio-signal generator (two audio signals may be used), see Comments

Test Setup:

See Procedure.

Comments:

The best way to check a single-sideband (SSB), class B amplifier for linearity and peak envelope power (PEP) output is with a scope. It is also a considerable help to have two audio tones available. But probably the most popular way to check the system is with a single audio oscillator plus some carrier. To add the carrier needed for the second audio tone, all that is required is that you unbalance the carrier balance control in the SSB rig. The PEP measurement described here does not meet FCC requirements. You should consult your local FCC Field Office for information pertaining to the Commission's rules and regulations. However, FCC regulations require that the transmitter power be rated in terms of the DC input to the final stage.

Procedure:

Step 1. Place a dummy antenna on the SBB rig and couple the vertical deflection plates of the scope's CRT to the transceiver output. Or, as an alternative, use an RF probe and detector and couple directly to the scope's vertical amplifier input. To make the coupling, you can use either magnetic coupling by using an RF pickup circuit (a couple of loops of wire placed near the transmitter output and connected directly to the scope's CRT vertical deflection plates), or a modified coaxial connector coupling tee such as the one shown in Figure 10-6. The tee connector is installed in-line between the SSB transceiver and dummy antenna. The scope CRT vertical plates are capacitive coupled to the RF line, when connected to the straight adapter.

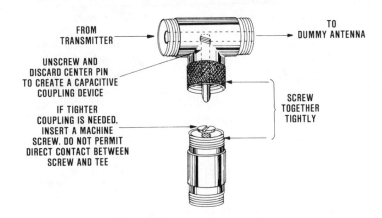

FROM TRANSMITTER

TO DUMMY ANTENNA

UNSCREW AND DISCARD CENTER PIN TO CREATE A CAPACITIVE COUPLING DEVICE

IF TIGHTER COUPLING IS NEEDED, INSERT A MACHINE SCREW. DO NOT PERMIT DIRECT CONTACT BETWEEN SCREW AND TEE

SCREW TOGETHER TIGHTLY

Figure 10-6: A capacitive coupler made using common coaxial connector components

Step 2. It may be necessary to adjust the size of the scope pattern by moving the RF pick-up coil, or adjusting the machine screw in the capacitive coupler shown in Figure 10-6. Set the scope internal sweep for about 30 Hz.

Step 3. Set the SSB rig for normal voice operation and apply two sine-wave audio signals about 1,000 Hz apart (for example, 600 kHz and 1.6 kHz, or 3 kHz and 4 kHz), to the transmitter's audio input circuit. To do this, you can parallel two audio-signal generators, or use one audio-signal generator and some carrier, as mentioned in

Comments. If you use this method, adjust the carrier unbalance control and signal generator output level control until you have equal amplitudes.

Step 4. Now, adjust the audio gain on the SSB rig and observe the pattern on the scope. You should see one of the patterns shown in Figure 10-7, depending on how you have the instrument controls adjusted and the equipment set up. The condition shown in (C) should not be possible unless you have a trouble, use high drive, and/or disable the A/C system found in most rigs.

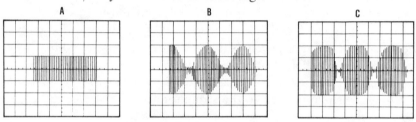

Figure 10-7: Pattern obtained by modulating an SSB transceiver by a single audio tone is shown at A. The pattern at B is obtained by modulating an SSB rig with a combination of two audio tones. Serious distortion and splatter will result with a transmitter adjusted to produce the pattern shown in C.

Step 5. Read the peak voltage (using an RF voltmeter is the easiest and most accurate way to do this). Next, convert your peak reading to rms by multiplying by 0.707. Then simply use the formula $P = E^2/R$. Simplified, all we have is this: $PEP = (E_{peak} \times 0.707)^2/R$, where R is the impedance of your dummy antenna.

10.6 A SIMPLE SSB MODULATION CHECK

Test Equipment:

Oscilloscope

Test Setup:

See Procedure.

Comments:

A measurement of percent of modulation for an SSB transmitter has no real meaning in reference to a measurement of an AM transmitter percent of modulation. In fact, the meter readings cannot be relied on for assurance that the transmitter is being properly modulated. The only way to be sure that the SSB rig is being operated within linearity limits is to use a scope and modulate the transceiver with your voice. However, the method explained under the heading SSB Power Output Measurement (Test 10.5), is a very practical way to check modulation patterns. The limit of modulation on the SSB transmitter is set by the point at which *your voice* causes flattening to begin at maximum modulation. Figure 10-8 is an illustration of a typical SSB voice modulated signal. It is important to realize that the modulation pattern shown is only instantaneous and would be continuously varying in amplitude and pattern during normal speech.

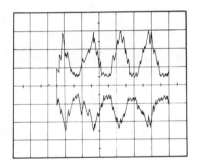

Figure 10-8: An example of what type pattern may be seen on an oscilloscope during voice modulation of an SSB transmitter

Procedure:

Step 1. Couple the vertical input of the scope to the transmitter RF output stage. The simplest coupling method is to use a piece of hook-up wire and form a couple of turns at one end and connect the other two ends to the scope vertical input terminals. Place the pick-up coil near the transmitter's output stage and use the loosest coupling that will give a convenient height to the scope pattern.

Step 2. Set the scope to internal sweep. You'll need a slow sweep rate.

Step 3. Now, speak into the SSB rig microphone and adjust the signal level for maximum possible height ... one that doesn't show overloading. You want the best looking pattern with no clipping, because clipping off the tops (flat topping) can cause serious distortion and considerable increase in spurious frequencies, causing splatter (see Figure 10-7 C).

Step 4. After you have established the maximum possible height on the scope pattern without flattening of the peaks, mark this level on the face of the scope with a grease pencil, if you want to leave the scope hooked up during normal operation. Keeping the peaks just below your grease pencil marks will help assure you that you are not causing splatter and your signal will be free of distortion.

10.7   FM PERCENT-OF-MODULATION MEASUREMENT

Test Equipment:

AF signal generator, deviation meter, 50 ohm dummy antenna, and coaxial connecting cables

Test Setup:

See Figure 10-9.

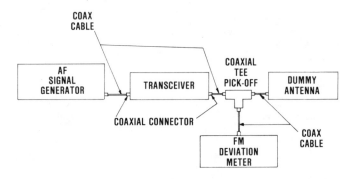

Figure 10-9: Test setup for making an FM modulation measurement

Comments:

Refer to SSB Power Output Measurement (Test 10.5) for construction details of a coaxial tee pick-off attenuator. A simple deviation-meter schematic that you can build is given in the 1977 Radio Amateur's Handbook, published by the American Radio Relay League.

Procedure:

Step 1. Assemble all equipment, as shown in Figure 10-9. Then load the transmitter into the dummy load.

Step 2. Set the AF signal generator to 1,000 Hz and adjust the FM deviation meter according to the manufacturer's instruction manual.

Step 3. Adjust your AF signal generator output level until you have a frequency swing of a predetermined value. The exact value of frequency swing for 100% modulation will depend on what is authorized by the FCC. For example, in FM broadcasting it is ±75 kHz and, if you're servicing certain mobile radios, it's ±5 kHz (ignoring permitted tolerances in both cases).

Step 4. To determine the percent of modulation, divide the reading on the deviation meter by the maximum authorized swing and multiply by 100. As an illustration, let's say that you're checking a mobile radio where the maximum swing permitted is ±5 kHz. Now, assume you read ±2.5 kHz on the deviation meter. Next, 2.5 kHz/5 kHz × 100 =50% modulation. To find the deviation ratio, simply divide the maximum authorized RF frequency swing by the maximum audio frequency that is authorized for the purpose of modulating the FM transmitter. Using another mobile radio example, 5,000/3,000 = 1.6 deviation ratio.

Step 5. Increase the AF signal generator output level while observing the deviation meter. If your reading exceeds the authorized swing (for example, ±5 kHz for mobile radio), the transmitter's deviation control must be re-adjusted until it prevents overmodulation.

10.8   TRANSMITTER FREQUENCY MEASUREMENTS

ABSORPTION WAVEMETER METHOD

Test Equipment:

Homemade absorption wavemeter shown in Figure 10-10, mounted in a metal box

Test Setup:

See Procedure.

Comments:

If you need to make a frequency measurement where accuracy

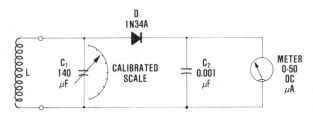

Figure 10-10: Schematic for a simple absorption type frequency meter that, with suitable coils, will cover a broad range of frequencies

is important — such as complying with the FCC — *this method is not recommended.* However, for rough checks of receiving equipment, transmitter tune-ups, and many other shop checks where accuracy is not of prime importance, this simple instrument and procedure will speed up troubleshooting and quickly isolate the fundamental frequency because it is very insensitive to harmonics (when loosely coupled to a tuned circuit). Two formulas that can be used in the design and construction of an air core coil are given below.

$$L = \frac{r^2n^2}{9r + 10D} \quad \text{and} \quad n = \frac{\sqrt{L(9r + 10D)}}{r^2}$$

where **L** is in microhenrys, **r** is radius of the coil in inches, **D** is length of coil in inches, and **n** is the number of turns. Table 10-1 lists the construction details for four coils that can be used with the wavemeter shown in Figure 10-10, which will provide a frequency span of approximately 1 to 150 MHz. Other coils can be designed using the previously mentioned formulas.

APPROX. FREQ RANGE	VARIABLE CAPACITOR C = 140 pF
(A) 1-4 MHz	72 TURNS NO.32 INSULATED WIRE CLOSE WOUND ON 1" DIAMETER FORM
(B) 4-12 MHz	21 TURNS NO. 22 INSULATED WIRE CLOSE WOUND ON 1" DIAMETER FORM
(C) 12-40 MHz	6 TURNS NO.22 INSULATED WIRE ON 1" DIAMETER FORM. ADJUST TO LENGTH OF 3/8.
(D) 40-150 MHz	1 SHORT PIECE OF NO. 16 BARE COPPER WIRE. FORM A HAIRPIN LOOP WITH 1/2" BETWEEN STRAIGHT SIDES. MAKE TOTAL LENGTH, INCLUDING BEND, 2"

Table 10-1: Construction details for four coils needed to provide an approximate frequency span of 1 to 150 MHz, when used with the wavemeter shown in Figure 10-10

You'll have to make a dial for the frequency meter and connect it to the variable capacitor. To prevent over-crowding of a single dial (if you want the full frequency coverage shown in Table 10-1), it's best to make a dial for each of the ranges shown. You can calibrate the instrument by using a signal generator or any other accurate source of RF energy. When you finish the circuit, place it in a metal box and take care during use that body capacitance does not affect your measurements.

Procedure:

Step 1. Select a coil that covers the frequency range of the equipment under test, connect it to the wavemeter tuned circuit and place a calibrated dial on the capacitor shaft. Check your instrument calibration with a known-to-be-good signal source, if it hasn't been done recently.

Step 2. Energize the equipment under test and *loosely* couple the wavemeter coil to it. Tune the capacitor for resonance at the unknown frequency. You'll see peak deflection on the meter at resonance. Using very loose coupling insures that you are tuned to the fundamental frequency at a *maximum peak* (you'll probably find weaker peaks).

Step 3. After you are sure that you have the peak reading on the meter (you may have to vary the coupling and retune the capacitor a couple of times), simply read the unknown frequency off the wavemeter homemade dial.

FREQUENCY COUNTER METHOD

Test Equipment:

Frequency counter or frequency meter, coaxial tee pick-off (see SSB Power Output Measurement, Test 10.5, for construction details), 50-ohm dummy antenna, and connecting coaxial cables

Test Setup:

Connect the equipment as shown in Figure 10-11.

Comments:

Transmitter frequency is best obtained by direct measurement. Usually this is done by using a digital frequency counter. Typically, a low-cost counter will have an accuracy of 3 ppm at 25° C (this turns out

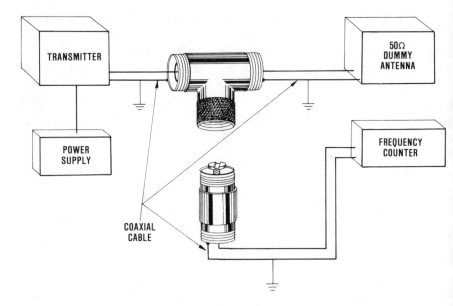

Figure 10-11: Test setup for making a transmitter frequency measurement
using a frequency counter

to be less than 30 Hz at 10 MHz), while more expensive ones (over $200) have 1 Hz resolution at 30 MHz. When checking some transmitters, you'll have to use a counter that operates at much higher frequencies. One counter that will check higher frequencies is the Motorola S-1324. This digital instrument provides continuous frequency coverage from 50 Hz to 525 MHz. Its specifications include resolution to 0.1 Hz and an oscillator stability of 3 parts in 10^9.

There are numerous other frequency counters on the market, but the important point is that you must determine what frequency range and accuracy you need for servicing a particular transmitter. To do this, you must consult FCC rules and regulations (your local Federal Communications Commission Field Office can help you determine what government publications you need) for the service you are interested in (Public Safety Radio Service, Broadcast Service, etc.).

Procedure:

Step 1. Connect the equipment as shown in Figure 10-11 and wait until the radio is operating within specified temperature limits. Be sure the temperature is correct because any adjustments to the assigned frequency may result in an out-of-tolerance operation when the set is operated at the specified temperature.

Step 2. Key the transmitter and read the frequency off the frequency counter.

Step 3. Adjust to the assigned frequency, if necessary. Record the *actual* frequency in the log. Do this on all channels. *Note:* Although this adjustment can be done by anyone, it's *important* that the transmitter be checked by a person holding a valid FCC First or Second Class Commercial License before it is connected to an antenna and used to radiate RF energy into the air (except for amateur radio operators, where an Amateur License is required).

Chapter 11

Tests and Measurements for Antenna
Systems and Transmission Lines

In this chapter you'll find that stress is placed on making practical antenna measurements and tests with inexpensive equipment. Like most of the chapters in this book, this one has selected pieces of electronic test instruments that are easily home-built, and construction of this gear is described, or information is included that tells you where to find a schematic and parts list. As an example, you'll see how to make a quick systems test of a CB antenna — using a handy, homemade portable test antenna *that you can trust*. This test will get your troubleshooting job off in the right direction — toward fast, accurate diagnosis and easy repair.

You'll find the step-by-step instructions used in every measurement will help you speed through such testing procedures as determining antenna impedance, transmission line velocity factor, antenna front-to-back ratio, and more. Each one clearly maps out instrument setup and connections for easier measurements. Just go to the section that describes the exact test or measurement procedure for your particular need. Use the Table of Contents or Index to quickly find the one you're after. Each test is totally complete — or makes reference to another specific test, by test name and/or number. You'll find that the working illustrations and comments include all the information you need to make the test setup (or building instructions, if it's a homemade instrument), plus any additional shop tips that could help you skillfully conduct a successful and accurate diagnosis.

11.1 MOBILE AM-FM RECEIVER
ANTENNA CIRCUIT TEST

Test Equipment:

None

Test Setup:

Extend the antenna to maximum receiving height.

Procedure:

Step 1. Turn on the receiver and tune it to a very weak station, or noise, in the frequency range of 800 kHz to 1400 kHz (AM broadcast band).

Step 2. Next, adjust the antenna trimmer capacitor for maximum signal output. You'll frequently find that there is a small hole on the receiver case, near the antenna jack, that gives access to the antenna capacitor for the purpose of peaking. Figure 11-1 shows a typical electrical diagram of the antenna system.

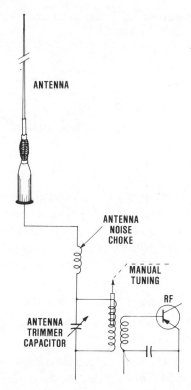

Figure 11-1: Mobile antenna circuit diagram showing antenna trimmer capacitor connections

Step 3. If you can't find a definite peak, it's an indication there is a defect in the antenna system. Your next step is to see Section 11.2, Antenna System Leakage Test.

11.2 ANTENNA SYSTEM LEAKAGE TEST

Test Equipment:

Ohmmeter and jumper lead

Test Setup:

See Figure 11-2.

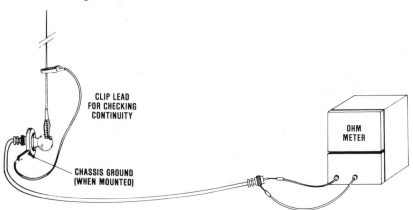

Figure 11-2: Test setup for testing an antenna system's leakage

Comments:

This test can be used to check out all types of antennas; mobile whip, vertical ground plane, TV, simple dipoles, etc. However, it cannot be used to check antennas using shunt feed or coupling networks that use impedance matching networks that present a DC short (for instance, the "delta" matching system).

Procedure:

Step 1. Disconnect the antenna transmission line from the transmitter or receiver.

Step 2. Connect a jumper lead across the antenna feed points, or between the whip and automobile chassis ground (rooftop, trunk lid, etc.) as shown in Figure 11-2.

Step 3. Measure the leakage resistance of the antenna system with the ohmmeter. To do this, connect one ohmmeter lead to the center conductor of the coax (if the system uses a coax lead-in), and the other to the outer shield connector. In the case of unshielded 300-ohm

TV lead-in, simply connect the leads across the two transmission line electrical conductors. The meter should indicate an open circuit. If leakage is detected, clean and check the insulator or, if you're checking a TV antenna system, check and/or replace the lead-in cable (assuming the antenna checks out okay).

11.3 HALF-WAVE DIPOLE ANTENNA TUNING TEST

Test Equipment:

Dip meter

Test Setup:

See Figure 11-3.

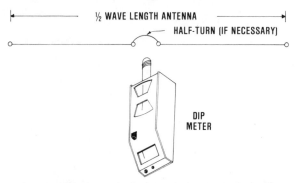

Figure 11-3: Illustration of how to couple a dip meter to a half-wave dipole antenna for the purpose of determining its loading frequency

Procedure:

Step 1. Place the appropriate plug-in coil in the dip meter and set it to operate in the frequency range you expect the antenna to operate.

Step 2. Next, loosely couple the meter coil to the center of the dipole, using a short piece of hook-up wire to create about one-half turn, if necessary (see Figure 11-3).

Step 3. Tune the dip meter for maximum dip and then read the frequency off the dip meter dial. *Note:* Generally, the accuracy of a dip meter is not exceptionally good.

Step 4. To improve the accuracy of your check, compare the dip meter to a calibrated communication receiver such as the ones used by

SWL's, amateur radio operators, etc. *Note:* Be sure all guy wires and supporting structures are non-metallic (use plastic lines). Also, other antennas will have a pronounced effect on any antenna you're testing, if they are closer than one wavelength to the antenna under test.

11.4 ANTENNA INPUT IMPEDANCE MEASUREMENT

Test Equipment:

Dip meter, impedance bridge and microammeter

Test Setup:

See Figure 11-4.

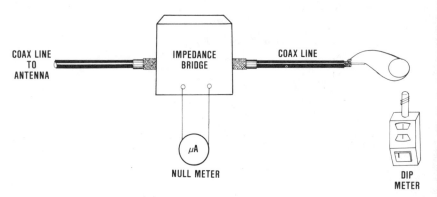

Figure 11-4: Test setup for measuring the input impedance of an antenna

Comments:

There are several methods of measuring an impedance that is composed of both resistance and reactance. For instance, you can use a vector impedance meter such as produced by Hewlett-Packard Co., of Palo Alto, Calif., that has a readout in terms of polar coordinates (Z = R∠0°). That is to say, the readout is in terms of ohms and phase angle. Or, you can use the following procedure that has measured values that are of equivalent series form, R±jX.

The reason for choosing the following procedure (although it isn't as accurate as using expensive instruments) is that the schematic diagram and complete construction details for an RF impedance bridge you can build, are readily available in American Radio Relay League publications such as "ARRL Antenna" book or the "Radio

Amateur's Handbook," — that can be found in almost any public library. Incidentally, you'll also find construction details for a dip meter in the "Handbook."

Procedure:

Step 1. Although it can be done, in general, it's best not to use a transmitter to excite the antenna during this test because it's easy to burn up an RF bridge. Therefore, as a first step, disconnect the antenna from the transmitter.

Step 2. Connect the load input terminals of the RF bridge to the feed point of the antenna under test.

Step 3. Couple the dip meter to the impedance bridge, using the proper plug-in coil. Figure 11-5 shows a magnetic coupling coil and coax lead that you can make up, if needed.

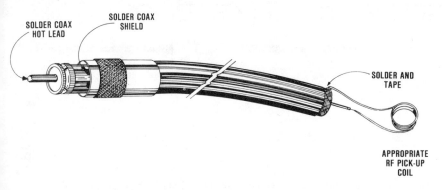

SOLDER COAX
HOT LEAD

SOLDER COAX
SHIELD

SOLDER AND
TAPE

APPROPRIATE
RF PICK-UP
COIL

Figure 11-5: RF pick-up link assembly that can be used between an RF bridge and dip meter being used to excite an antenna for test purposes

Step 4. Adjust both the dip meter and bridge, trying for a zero reading on the bridge null detector. If you can't get an absolute zero, it isn't all that important. However, it does mean the antenna isn't perfectly tuned. In other words, the antenna is acting as a slightly reactive load rather than a pure resistance. Trimming the antenna length or adding capacitance or inductance, as needed, should bring your reading closer to zero, if it's off considerably.

Step 5. After you have as near zero as possible, read the impedance of the antenna off the bridge indicating meter, and the antenna operating frequency off the dip meter.

11.5   TESTING A VERTICAL ANTENNA
WITH A BASE LOADING COIL

Test Equipment:

Dip meter, impedance bridge, and special connecting cables (see Procedure for construction details)

Test Setup:

See Figure 11-6.

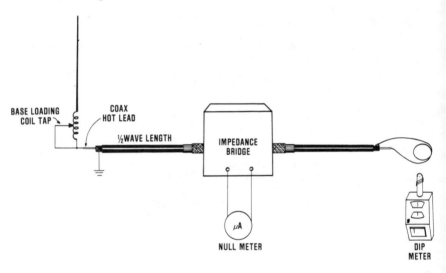

Figure 11-6:   Test setup for checking vertical antenna with base-loading coil

Comments:

Many commercially produced vertical antennas (especially for fixed station operation) have built-in resonant traps to make them broadband as well as making them appear electrically shorter. This test is not recommended for this type antenna because, in most designs, it does not have adjustable tuning coils. On the other hand, if you have a homemade vertical — such as a ground plane, etc. — using an adjustable base loading coil and aren't sure it is adjusted for best reception or to produce the best radiation pattern, this test will give you a rough idea of whether antenna adjustments are needed. See Comments for Test 11.4 for information pertaining to the construction of an RF bridge and dip meter.

Procedure:

Step 1. Make up a coaxial cable one-half wavelength long (the cable on the left in illustration 11-6) at the operating frequency of the antenna. You'll need only one coax connector at one end of the line. Connect the other end of the coax to the antenna, as shown, then make all other connections.

Step 2. Place the proper plug-in coil in the dip meter and adjust it to the approximate operating frequency of the antenna.

Step 3. Adjust the RF bridge, antenna base loading coil tap, and dip meter until you have the lowest reading possible on the null meter. Check the frequency by reading it off the dip meter. If it is too high, you need more inductance; too low, less inductance. Move your coil tap either up or down.

Step 4. The previous steps will let you know whether the antenna is properly tuned at the desired frequency, assuming the dip meter is accurate. To check the dip meter, you can zero beat it against a calibrated communications receiver such as used by radio amateurs and SWL's.

11.6   TRANSMISSION LINE VELOCITY
FACTOR MEASUREMENT

Test Equipment:

Dip meter

Test Setup:

See Procedure.

Comments:

See Test 11.4 comments for information about building a dip meter.

Procedure:

Step 1. First, you'll need to remove the insulation from a piece of coax and form a small loop (for coupling purposes) out of the inner conductor and then connect the inner conductor to the outside braid, as shown in Figure 11-7.

Step 2. Measure the exact physical length of the coax. Then use your calculator to determine the approximate *half-wave* resonant frequency of the line, using the formula:

$$\text{frequency (MHz)} = 468/\text{length (feet)}$$

Step 3. Next, set the dip meter to one-half your calculated frequency and couple it to the loop and adjust the tuning dial for a dip on the indicating meter. Now, simply use your calculator with the following formula to find the velocity factor.

$$\text{velocity factor} = \frac{\text{(length in feet) (dip meter frequency in MHz)}}{246}$$

The formula is based on the formula for a quarter-wave stub, therefore the figure for the dip meter frequency is one-half frequency of a half-wave line. The reason for this is that it is difficult to couple a dip meter to a coaxial cable with any accuracy. Using this method, will help the situation.

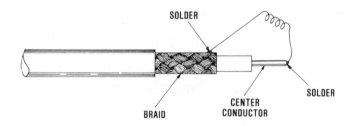

SOLDER

SOLDER

CENTER
CONDUCTOR

BRAID

Figure 11-7: Homemade coupling coil for measuring transmission line velocity factor

11.7 ANTENNA FRONT-TO-BACK
RATIO MEASUREMENTS

FIELD-STRENGTH METER METHOD

Test Equipment:

Field-strength meter

Test Setup:

See Comments and Procedure.

Comments:

Whenever antennas are being checked, they should be installed at the proper distance above ground, and preferably in an open field. Ideally, antennas should be at least one wavelength above ground (for example, 36 ft. at CB frequency). When testing, install both antennas at the same height and, if guy wires are used, they should be non-metallic (plastic lines).

Procedure:

Step 1. Place the field-strength meter at least three wavelengths away from the directional antenna under test. Adjust the meter until its antenna has the same polarization as the antenna under test. *Note:* Complaints of poor front-to-back ratio can usually be traced to reflections from other antennas, buildings, or power lines, so be sure you are clear of such obstacles.

Step 2. Adjust the directional antenna under test so that it points exactly at the antenna of the field-strength meter. Energize the transmitter feeding the test antenna and then note the reading on the field-strength meter (E_1 in the formula in Step 4). Turn off the transmitter.

Step 3. Next, adjust the directional antenna to point the opposite direction and turn on the transmitter and take a field-strength reading (E_2 in the formula in Step 4).

Step 4. Now, because a field-strength reading is a voltage measurement, you can convert to dB's by using the familiar formula:

$$dB = 20 \log_{10} (E_1/E_2)$$

where E_1 is your first field-strength reading (Step 2) and E_2 is your second field-strength reading (Step 3).

Step 5. Once you've done Step 4, your answer is the front-to-back ratio of the antenna in dB's. One final word of caution. If you are using a homemade field-strength meter, be sure that the antenna lead-in of the meter does not pick up RF energy from the test antenna because this will cause your measurement to be erroneous. Also, it's best to use a coaxial line to feed the test antenna, if possible, so you won't have transmission line radiation. This will also cause your readings to be incorrect.

S-METER METHOD

Test Equipment:

Receiver with an S-Meter

Test Setup:

See Field-Strength Meter Method Comments and Procedure in this section.

Comments:

Depending upon the transceiver, an S-meter is calibrated so that one S-unit is equal to 6 dB. Therefore, when using an S-meter to perform a front-to-back ratio measurement, you'll find that an antenna has 1 S-unit gain between the front-to-back measurement and will also have a front-to-back ratio of 6 dB. *Caution:* Some S-meters are calibrated at only 3 dB per S-unit, and others at 3 or 4 at the low end and 6 or 7 at the top of the scale. *If in doubt, consult the manufacturer's service information.*

Procedure:

Step 1. Place a receiver (with S-meter) and its antenna at least three wavelengths away from the directional antenna under test. See Step 1 in the previous method.

Step 2. Same as Step 2 of the previous method except that you note the S-meter reading, using this method.

Step 3. This step is the same as Step 3 in the previous method except, again read the S-meter.

Step 4. Note the gain in S-units between Step 2 and 3. If 1 S-unit equals 6 dB, the gain in S-units multiplied by 6 will be the front-to-back ratio in dB's of the antenna under test. As explained under Comments, you can run into other methods of calibration, but the basic procedure is still the same.

11.8 TRANSMISSION LINE CHARACTERISTIC IMPEDANCE MEASUREMENT

Test Equipment:

VOM, impedance bridge, and dip meter

Test Setup:

See Figure 11-8.

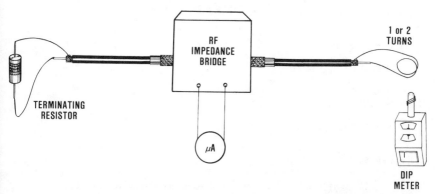

Figure 11-8: Test setup for measuring the characteristic impedance of a transmission line

Comments:

See Comments, Test 11.4, for information about building an RF bridge and dip meter.

Procedure:

Step 1. Connect your equipment as shown in Figure 11-8. The resistor you use as a terminating resistor (load resistor) can be any value, to start. However, it should be a carbon type (i.e., non-reactive), having a value that you think is about the same as the characteristic impedance of the transmission line under test. The impedance of two-wire lines ranges from 100 to 300 ohms and the most often encountered coaxial line impedances are 50 and 75 ohms.

Step 2. Couple the dip meter to the test cable coil, as shown, and set it to the desired frequency. If you have selected a resistor of the same value as the line's characteristic impedance, you'll be able to adjust the bridge for a zero reading on the impedance bridge.

Step 3. If you can't adjust the null meter to zero, try another resistor and again attempt to bring the null meter to zero reading.

Step 4. When you have a zero reading on the bridge, try setting different frequencies on the dip meter. The null meter should remain at zero with changes of frequency.

Step 5. Measure the value of your final resistor and this is the characteristic impedance of the line under test.

11.9   SWR MEASUREMENT

Test Equipment:

In-line SWR meter

Test Setup:

See Procedure.

Comments:

There are several popular low-cost directional coupler type SWR meters sold in kit form and already assembled (for example, Heathkit and Radio Shack). However, it's very easy to make a home-built one. You'll find schematic diagrams in many publications written for amateur radio operators. Quite a few of them are designed around the schematic diagram shown in Figure 11-9.

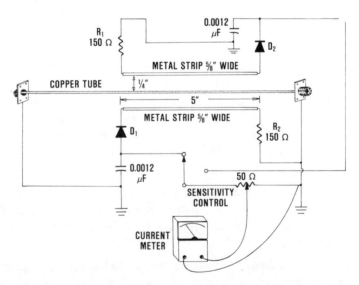

Figure 11-9:   Typical in-line SWR tester design similar to the ones used by electronics manufacturers and home builders. With components shown, the instrument works in a 52 ohm transmission line at a power of about 80 watts or more.

The current meter shown in Figure 11-9 is usually a 0-100 microammeter. However, it can be a milliammeter if the transmitter output is strong enough. Just about any radio frequency diodes can be used for D_1 and D_2. For example, a matched pair of 1N34A's will work well if the maximum forward current does not exceed 5 mA.

Procedure:

Step 1. Place the pickup section of the SWR meter between the transmitter and antenna feed line. It is important that the impedance of the SWR meter matches the impedance of the transmission line. In most designs like the one shown in Figure 11-9, you'll find that the impedance of the instrument can be changed by using different values of resistance for the two resistors labeled R_1 and R_2. For example, changing the resistor's value to 100 ohms instead of the 150 ohms shown will change the impedance of the unit to 75 ohms. See the manufacturer's instructions for other instruments of this type, to determine the value needed.

Step 2. Energize the transmitter and check that it is tuned to the correct operating frequency—peaking the tuning, if necessary.

Step 3. Check to be sure the SWR meter function switch is in the *forward position* and adjust the sensitivity control for a full-scale reading (if possible).

Step 4. Without making any other changes, place the function switch to the *reflect* position. If you're using a reflected power meter— such as Heathkit—simply read the SWR. But, if you're using a 0-100 μA meter in a homemade rig, you'll have to use the formula:

$$SWR = \frac{10 + \text{reflected reading}}{10 - \text{reflected reading}}$$

where 10 is full-scale reading. In other words, you're reading 100 μA as 10. For example, if you read 3 in the reflect position,

$$SWR = (10+3) \,/\, (10-3) = \text{about } 1.85 \text{ to } 1$$

11.10   PORTABLE CB TEST ANTENNA

Test Equipment:

Homemade coaxial antenna

Test Setup:

See Procedure.

Comments:

The quality of a coax and connectors, and especially the soldering of coax to the connectors can affect SWR and gain. If you suspect a CB antenna system is using a poor grade of coax (this can cause a loss of 2 or 3 dB), or the system has poor soldering, a quick check can be run, using the following portable antenna. Figure 11-10 shows the simple construction details.

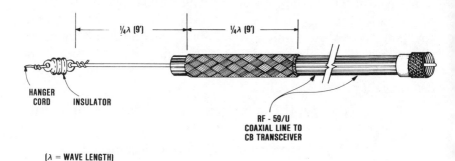

Figure 11-10: Construction details of a CB test antenna that can be taken almost anywhere. Cut off the outer cable insulation for a length of 9 feet and fold the shielded braid back over the cable, as shown.

Procedure:

Step 1. Disconnect the CBer's antenna system from his transceiver.

Step 2. Connect your homemade antenna to his transceiver.

Step 3. String up the test antenna, attaching the insulator to some high point. Keep the test antenna as far as possible from other antennas, power lines, or any other material of a conductive nature. Ideally, nothing should be closer to your test antenna than 36 ft. (1 wavelength).

Step 4. Energize the transmitter and try to contact another CB station. If you are successful, you can be sure that the original antenna system needs replacement or repair.

Chapter 12

Automotive Electrical Systems
Tests and Measurements

Here are step-by-step procedures that show you how to expertly test the electrical systems used in all types of gasoline-driven engines. For example, how to test emergency vehicles, electric power generators, chain saws, lawn mowers and, not the least important, your car, are covered. The drawings you'll find in the following pages can be clearly visualized without having to search for and piece together several schematic diagrams. You'll find that the illustrations and step-by-step procedures are so well presented that they will eliminate all frustrations so you'll know exactly what steps to take to complete your testing job quickly and easily.

Using this chapter, you'll find that you can immediately make the necessary test hook-up and then follow the simple test technique with an absolute minimum of time and effort. For example, did you know that you can check for a fouled or cracked spark plug without even removing the plug from the engine, using nothing but a VOM? Or, as another example, did you know that if a mechanic substitutes one wrong ignition cable during a tune-up, it can almost wipe out communications with a CB rig or any other type radio receiver? Test 12.2 tells you how to check out an interference problem like this quickly and, again, using nothing but a VOM. You'll also find solid state ignition systems tests included because transistorized ignition systems are a fact of life in the engines of today. Each test presents a comprehensive approach to servicing and troubleshooting an ignition system that any electronics technician can perform with an absolute minimum of low-cost test gear—and you don't have to be an automotive mechanic to enjoy doing them.

12.1   MOBILE RIG NOISE INTERFERENCE TEST

Test Equipment:

None

Test Setup:

None

Comments:

All capacitors listed in the following steps should be special automotive types such as can be purchased from such companies as Radio Shack, etc. See Figure 12-1.

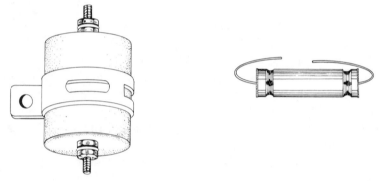

Figure 12-1: Two typical automotive type coaxial capacitors

Procedure:

Step 1. Start the car engine and turn on the radio receiver (AM, FM, CB).

Step 2. If you hear a popping static at a regular pulsed rate, that varies with engine speed, it's an indication that the resistance ignition wire (used in most American cars) may be defective. It's also possible that the antenna ground could be at fault (see Test 11.2). In most cases (except solid state ignition systems, in which case consult the car manufacturer's local service shop), you can try placing a 0.5 μF capacitor between the battery terminal of the ignition coil and ground, and it should help.

Step 3. If you hear a high pitched whistle, it's probably caused by the alternator. Placing a 0.1 to 0.5 μF capacitor and/or an L-section filter in the *receiver power lead* will usually clear up the problem. *Note:* Generally, it is best not to place an L-section filter in the power line feeding the transmitter, due to the high current drain. Auto noise eliminator kits that will handle high current are available from electronics supply companies (for example, Radio Shack).

Step 4. **If you hear a frying sound,** it indicates a possibly defective regulator. In some systems, you can try placing a 0.1 μF capacitor on the input and output leads of the regulator. Of course, this isn't practical when the voltage regulator is built into the alternator, as it is in many alternators such as some that are used in vehicles manufactured by the GM company.

Step 5. **If you hear a slow steady popping sound** that does not vary with a change in engine speed, it may be the gas gauge sender unit. Place a 0.5 μF capacitor between the wire coming from the gas tank (as close to the tank, as practical) and ground.

Step 6. Step on the brake pedal several times and listen for a popping sound. In this case, you're checking the brake light switch for interference. If you find interference, place a 0.5 μF capacitor across the switch.

Step 7. If all the previous steps do not completely eliminate all interference, place a 0.5 μF capacitor across (or in-line with) the heater-motor windings, horn switch, and any other switches, motors, etc. However, *do not* bypass the field winding in older car generators.

12.2 RESISTANCE IGNITION WIRE TEST

Test Equipment:
> Ohmmeter

Test Setup:
> See Procedure.

Comments:
> If you are having RF interference problems, don't overlook the possibility that a mechanic may have substituted straight copper high-voltage wires for the resistance wires that were originally used by the manufacturer, especially, if the interference problem pops up right after a tune-up job. Should you suspect that this has happened, simply perform the next two steps and you'll remove all doubts.

Procedure:
> *Step 1.* First, measure the resistance of the wire that runs from

the high-voltage terminal on the coil to the center terminal on the distributor. You should measure several thousand ohms. If not, replace it with the manufacturer's recommended wire. Incidentally, if the mechanic made the mistake, he, or the garage he works for, should replace it—*at no cost.*

Step 2. Next, measure the resistance of each of the spark plug wires. You'll read a high resistance in each measurement (about 5,000 ohms per foot). If you don't, replace the particular wire at fault. By the way, another sign that you're having trouble with the high-voltage wiring is rough engine idle. In fact, if there is radio interference and rough idling, it's a sure sign of either a wrong cable (or cables), or simply old-age-deteriorated ignition wiring.

12.3 SPARK PLUG CABLE TEST

Test Equipment:
Ohmmeter

Test Setup:
See Procedure

Comments:
If a vehicle engine runs rough (misses) when the engine is idling, it's very possible there are one or more defective spark plug wires. You can use the following procedure to quickly locate the defective cable (also see Tests 12.1 and 12.2). Be suspicious of any vehicle that has just had a solid state ignition installed, or a tune-up job, because rough handling of spark plug wires can cause open circuits.

Procedure:
Step 1. Start the car and let it run at idle speed.

Step 2. Remove one spark plug cable at a time and listen for any changes in sound coming from the engine. In other words, note if removing each of the cables (one-by-one) causes the engine to run even rougher than it did. What you are looking for is a cable that causes no difference, connected or disconnected.

Step 3. Once you have located the cable that *made no change,* turn off the engine and remove that cable from the electrical system.

Step 4. Next, measure the resistance of the cable with an

ohmmeter. If the cable is a resistance type, you should measure several thousand ohms, but not infinity. A reading of an infinite resistance indicates the cable is open and it should be replaced.

Step 5. Your next step is to flex the cable. If you observe an erratic indication (fluctuation of readings) on the ohmmeter, replace the cable because it has an internal short.

12.4 SPARK PLUG TEST

Test Equipment:
   VOM

Test Setup:
   See Procedure.

Comments:
   You can use your ohmmeter to check spark plugs — without removing them from the engine. However, this test will not tell you if a plug has burned out electrodes, or if the gap is set incorrectly. What it will tell you is whether the points are fouled, or if there is a current leakage path in the porcelain insulator. That is to say, it will find a cracked plug.

Procedure:
   Step 1. Set your VOM function switch to read resistance. Set the range switch to a high resistance range.

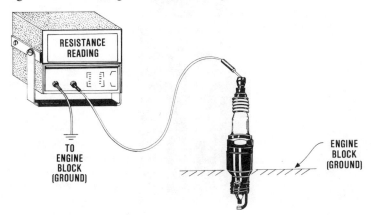

Figure 12-2: Ohmmeter connections for checking for a fouled or cracked
            spark plug

Step 2. Remove the ignition cable from the end of one plug.

Step 3. Connect the ohmmeter leads between the spark plug and engine block as shown in Figure 12-2.

Step 4. Measure the resistance. You should read an open circuit (infinite resistance). If you don't, pull the plug and measure the resistance between the points and case. A low resistance reading indicates a fouled or shorted plug. Clean or replace it.

12.5   IGNITION COIL PRIMARY TESTS

VOLTAGE METHOD

Test Equipment:

>   VOM

Test Setup:

>   See Figure 12-3.

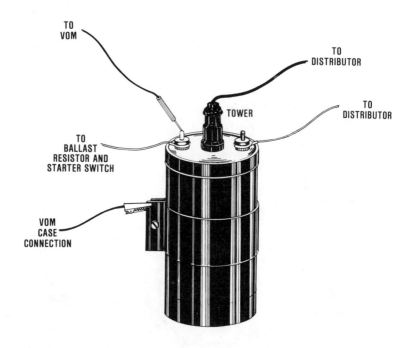

Figure 12-3: Voltmeter connections for measuring an ignition coil's primary voltage

Comments:

A capacitive-discharge (CD) ignition system has approximately 400 VDC delivered to the coil primary. Therefore, *take care* and refer to the manufacturer's service notes when checking a CD system.

Procedure:

Step 1. Set the VOM to read the full battery voltage of the vehicle or emergency generator you are testing.

Step 2. Connect the VOM test leads as shown in Figure 12-3.

Step 3. Turn the ignition switch on and crank the engine for about five seconds and observe the voltmeter reading at the same time. Your voltmeter reading should average not less than 7½ volts. If your reading is this high or better, the battery, starter, connecting cables, ignition switch, and all other circuits to the coil are good (this includes solid state components, if there are any in the circuit).

Step 4. If you get a zero or low voltage reading, switch to continuity checks using the ohmmeter. Check the coil's primary winding, cable resistance (you should see several thousand ohms when measuring HV cables, in most cases), and the ballast resistor (typically, you should measure a resistance of 0.5 to 2 ohms), plus all other items that are listed in Step 3.

OHMMETER METHOD

Test Equipment:

Ohmmeter

Test Setup:

See Figure 12-4.

Procedure:

Step 1. Disconnect the coil's primary leads from their connecting terminals and pull the secondary lead out of the coil. In other words, disconnect the coil from the vehicle's electrical system.

Step 2. Connect the ohmmeter leads to the primary side of the coil, as shown in Figure 12-4. An infinite resistance reading on the ohmmeter indicates an open coil winding.

Step 3. Move the ohmmeter's negative lead to the case of the coil to check to see if the coil is shorted to ground. Set the ohmmeter range switch to a high ohms position. You will read an infinite resistance (an open circuit). If you read a very low resistance, the coil is shorted and should be replaced.

12.6 IGNITION COIL SECONDARY TEST

Test Equipment:
Ohmmeter

Test Setup:
See Figure 12-4 and Step 2 in the following Procedure.

Procedure:
Step 1. Disconnect the distributor wire and coil terminal cable of the coil.

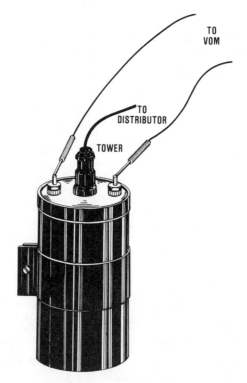

Figure 12-4: Ohmmeter connections for using the ohmmeter method to check an ignition coil's primary resistance.

Step 2. Set the ohmmeter range switch to read a low value resistance (about 10 k ohms) and refer to Figure 12-4. You can connect one of the ohmmeter leads to either of the coil's screw terminals (found on both sides of the coil and shown in Figure 12-4). Stick the other lead into the coil tower. If your leads are terminated with alligator clips, insert a screw driver (or something similar) into the coil tower and connect the alligator clip to the prod.

Step 3. You should read something like 5 or 10 thousand ohms, depending on the coil's manufacturer. However, most coils will measure in this range, if they are good.

12.7   A QUICK DISTRIBUTOR TEST

Test Equipment:

Voltmeter

Test Setup:

See Procedure and Figure 12-5.

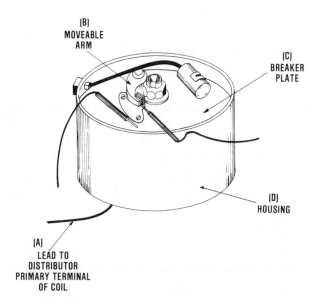

Figure 12-5:   Voltmeter connections for testing a distributor. See Procedure for connections at (A), (B), (C) and (D)

Comments

Excessive resistance within the distributor of an emergency electric power generator can seriously impair operations when emergencies arise in a community, because the generator cannot be started. This is particularly important to persons engaged in public service work. Of course, this is true for your car as well. The following simple procedure will quickly isolate the problem to the distributor if it's the culprit.

Procedure:

Step 1. Remove the distributor cap and crank the engine until the points are closed. Set the voltmeter to read on its lowest DC range.

Step 2. Connect the voltmeter to the distributor primary terminal of the coil (A) and to the movable arm of the breaker points (B). You should read a voltage not greater than 0.05 volts. In other words, using a low-cost voltmeter your reading probably will be zero, if the resistance is as low as it should be between these two points.

Step 3. Next, connect the voltmeter between the movable arm of the breaker points (B) and the breaker plate (C). Again, your reading should be as near zero as possible. Ideally, you shouldn't find more than a couple of tenths of a volt.

Step 4. Now connect the voltmeter between the breaker plate (C) and the distributor housing (D). As in all the previous Steps, your reading should be very low—not more than a few hundredths of a volt (about 0.05 volts).

Step 5. To make the last check, connect the voltmeter between the distributor housing (D) and engine block. If you read a voltage higher than about 0.05 volts, it's too much. The problem is that there is an excessive resistance path between the two units, probably due to corrosion, etc. Clean the unit and again make the measurement. Keep doing this until you read as near zero voltage as possible. Any reading close to 0.05 volts indicates a good electrical connection between the components.

12.8 CONDENSER (CAPACITOR) GROUND CONTINUITY TEST

Test Equipment:

VOM

Test Setup:

See Figure 12-6.

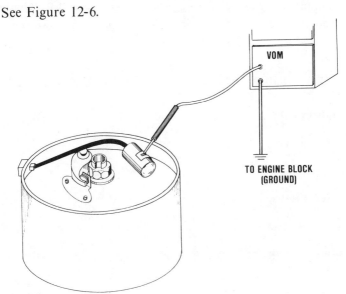

Figure 12-6: Ohmmeter connections for testing a condensor ground circuit

Comments:

This test should be made any time there is the least suspicion the suppression capacitor (condensor) ground circuit is defective. For example, if you're experiencing difficulty in starting the engine (or the points are excessively burned), it's quite probable that you have a grounding problem. Even a fraction of an ohm resistance between the capacitor case and ground can prevent proper operation of the ignition system.

Procedure:

Step 1. Remove the distributor cap and rotor (if necessary) and connect the ohmmeter as shown in Figure 12-6.

Step 2. Measure the resistance between the car ground (engine block, etc.) and the capacitor's metal case. It is essential that the capacitor makes a good ground connection through the capacitor clamp. Ideally, you should read zero resistance even using an expensive laboratory type ohmmeter. Practically speaking, replacing the ohmmeter with a voltmeter shouldn't produce a voltage reading over 0.05 volts. Which brings us right back to what was said before. You should read zero resistance in any practical situation. To make a measurement of the capacitor value, see Test 4.9 (Capacitance Measurements), or to make a quick, simple go-no-go test, see the next test.

12.9   QUICK OUT-OF-CIRCUIT CONDENSER TEST

Test Equipment:

> VOM

Test Setup:

> See Figure 12-7.

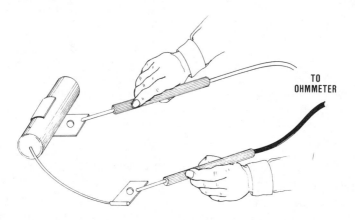

Figure 12-7:   Ohmmeter connections for testing an automotive condenser

Procedure:

Step 1. Set the VOM to read a high resistance and connect the test leads across the capacitor, as shown in Figure 12-7. In actual practice, it isn't really necessary to completely remove the capacitor from inside the distributor to check it. All that is required is that you disconnect one side of the condenser before using the ohmmeter.

Step 2. Note the reading on the ohmmeter. If the capacitor is shorted, of course you'll read zero resistance. On the other hand, if it is good, you'll see the meter pointer momentarily rise to a low value resistance and then fall back to a high value resistance and stay there.

12.10   IN-CIRCUIT ALTERNATOR DIODE TEST

Test Equipment:

Oscilloscope

Test Setup:

See Figure 12-8.

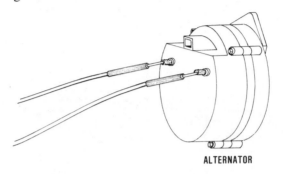

ALTERNATOR

Figure 12-8:   Scope lead connections for checking one type alternator diodes. When testing alternators that do not have connecting terminals as shown, it's necessary to connect the scope vertical input lead to the car's "hot" battery terminal.

Comments:

There are several ways to check alternator diodes. For example, use an ohmmeter (as explained in Chapter 1), or use a special purpose automotive tester. However, many electronics technicians have an ordinary scope in their shop that will not only check the diodes under *actual operation,* but will also check the alternator stator. Furthermore, the alternator doesn't have to be removed from the engine, which makes this way of testing very attractive. As a general rule, if you see any pattern on the scope except the one shown in Figure 12-9, the alternator stator or diodes are bad and you'll have to pull the unit for further tests, or take it to a local automotive mechanic for repair or replacement.

Procedure:

Step 1. Connect the scope vertical input lead to the battery terminal (usually a red cable) on the alternator. Connect the scope ground lead to either the engine block (ground) or to the alternator ground terminal (usually the black cable).

Step 2. Start the engine and let it run. Adjust the scope for a stationary pattern on the viewing screen. If the alternator is operating properly, you should see a DC output with a high frequency ripple voltage. Figure 12-9 shows an example of a pattern that is close to what you should expect to see when the alternator has no defects.

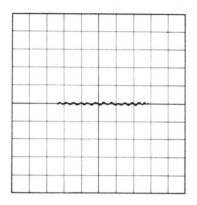

Figure 12-9: Scope pattern that should be seen when checking an alternator, if it is operating properly. The number of pulses seen will depend on the scope control setting.

Step 3. If you see a waveform that *resembles* a composite video signal like the one shown in Figure 12-10, it's an indication of a defective diode.

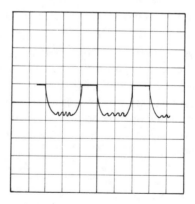

Figure 12-10: Alternator output waveform that will be seen on a scope if a diode is open. A shorted diode will produce essentially the same pattern except for less ripple between the square-wave parts of the pattern.

12.11   SOLID STATE IGNITION SYSTEM
SHORT CIRCUIT TEST

Test Equipment:

VOM

Test Setup:

See Figure 12-11 and Comments.

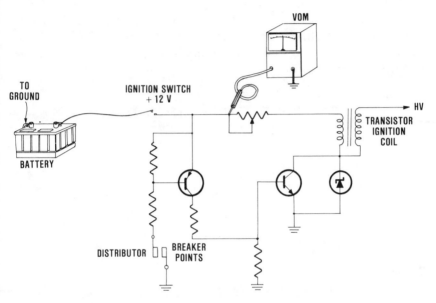

Figure 12-11: Simplified example of a transistor ignition circuit showing the voltmeter connection for a short circuit test. Although you'll find many different types of circuits, in general, you can always connect to the ignition switch wire for this test.

Comments:

Manufacturer's have designed a great many transistorized ignition systems so it isn't practical to try to explain them all. Fortunately, the testing and troubleshooting of the electronic module is basically the same as any other solid state device. However, as a general rule, the first problem is to find out if the trouble is in the transistorized module or in the associated components and wiring. *Note:* The voltmeter hook-up shown in Figure 12-11 will not work in all types of solid state ignition systems. Because of the variety in designs, it's always best to refer to the manufacturer's service notes.

Procedure:

Step 1. Remove the distributor cap so you can watch the distributor points. Energize the starter switch for short periods of time until you have the points in the open position.

Step 2. Connect a voltmeter with the positive lead to the ignition switch line (in many cases, this will be to the emitter of a transistor), as shown in Figure 12-11, and the negative lead to ground. Set the voltmeter to read more than 12 VDC (for example, the 15 volt range).

Step 3. Turn the ignition switch on *without cranking the engine.* With the switch on, you should read the full battery voltage. If your reading is low, or zero, either the battery is in poor condition or there is a short in the circuit.

12.12   SOLID STATE IGNITION SYSTEM
OPEN CIRCUIT TEST

Test Equipment:
VOM

Test Setup:
See Figure 12-11.

Procedure:

Step 1. The same as Test 12.11 except set the points so they are closed.

Step 2. Connect the voltmeter as shown in Figure 12-11.

Step 3. This step is the same as Step 3 in Test 12.11 except you *should not* read the full battery voltage. In this test, current should flow and the voltage should read about 6 or 7 volts. If you don't get a lower reading, look for a defective circuit.

12.13   DWELL EXTENDER TYPE SYSTEM TEST

Test Equipment:
VOM

Test Setup:
Connect the voltmeter positive lead to the negative terminal of

the ignition coil primary: connect the VOM negative lead to the car frame (ground). See Figure 12-12. Set the voltmeter to read about 12 VDC.

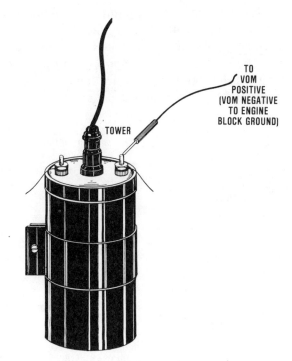

TO
VOM
POSITIVE
(VOM NEGATIVE
TO ENGINE
BLOCK GROUND)

TOWER

Figure 12-12: Voltmeter connections for dwell extender type system test

Comments:

Figure 12-13 is a schematic diagram of a silicon controlled rectifier (SCR) dwell extender, sold throughout this country. This test is checking the breaker points and electronics.

Procedure:

Step 1. Remove the distributor cap and energize the engine in short bursts until you have the breaker points set in the closed position.

Step 2. Connect the voltmeter as explained under Test Setup. Turn the ignition switch on and leave it on (without cranking the engine).

Step 3. You should now read a DC voltage very near the car battery voltage.

Step 4. Next, use your finger to open the breaker points. You should see the voltage reading fall to zero and immediately rise back up to the same reading you had in Step 3.

Step 5. Open and close the points several times and you should see a rise and fall in the voltage each time. *Note:* The rise and fall are very quick and it's possible that all you'll see is just a slight downward deflection of the reading each time you open the points. If you see a slight downward deflection, the dwell extender is good. If not, check the wiring, points, capacitor, and dwell extender.

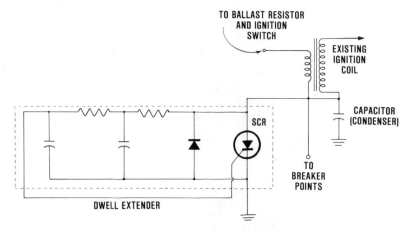

Figure 12-13: Simplified schematic diagram of an SCR dwell extender that can be checked with a voltmeter

12.14 SCR DWELL EXTENDER TEST

Test Equipment:

VOM

Test Setup:

See Procedure.

Procedure:

Step 1. Disconnect the SCR dwell extender from the distributor breaker points.

Step 2. Try to start the engine and see if it will run. If the car engine will start and run, the problem is in the dwell extender.

Step 3. As a general rule (after you've completed Step 2), it's best to first check a dwell extender for shorts with an ohmmeter. It isn't uncommon to find that the problem is a shorted SCR. See Test 1.4, Silicon Controlled Rectifier Test, in Chapter 1, for a complete SCR checkout procedure.

12.15 MAGNETO-TYPE CD IGNITION TESTS

Test Equipment:

See Procedure

Test Setup:

See Procedure:

Comments:

Figure 12-14 is a schematic diagram of a magneto-type capacitance discharge (CD) ignition system found in small gasoline-driven engines used to drive some emergency electric power generators and other devices such as snowmobiles, chain saws, and the like. The procedure given below tells you where to find the different tests needed to check out this type of ignition system.

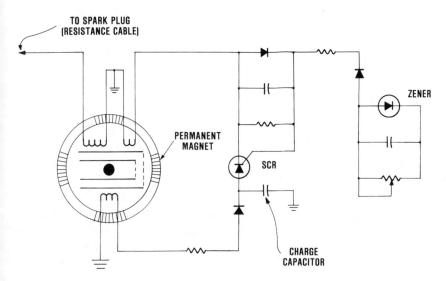

Figure 12-14: Schematic diagram of a CD ignition system used in some small gasoline-driven emergency generators

Procedure:

Step 1. Test the spark plug, cables and, *very important,* the ground lead, using a VOM, as explained in this chapter. Also, it's a good idea to use *sandpaper* and clean the outside edge of the permanent magnet rotor and the facing pick-up unit that houses the solid state components and coils, etc., if you encounter excessive dirt and rust on the rotor.

Step 2. Test the SCR, zener, and diode, using the procedures and equipment listed in Chapter 1. The other components are checked using standard troubleshooting procedures. *Note:* You may have up to 30,000 volts at the spark plug, its cable, and at the spark coil when you try to start the engine or when it's running. Also, you'll find a relatively high voltage across the charge capacitor, so be careful when testing in these circuits.

12.16   HORN ELECTRICAL SYSTEM TEST

Test Equipment:

VOM

Test Setup:

See Figure 12-15.

Procedure:

Step 1. Connect the positive lead of the voltmeter to the horn lead at the horn relay, as shown in Figure 12-15. Connect the negative lead to ground.

Step 2. Set the VOM function switch to read DC voltage and set the range switch to measure either 12 or 6 volts (it may be higher, see *Note*), depending on the car's electrical system voltage. *Note:* The voltage across the terminals of a vehicle battery varies widely from its 12 (or 6) volt level. For example, when fully charged a so-called *12 volt system* will have a battery voltage of 12.6 volts, but even this depends on the temperature. On cold days, it will be less than 12 volts. Furthermore, depending on the voltage regulator setting, with the engine running and the generator charging, you may find the voltage as high as 15 volts.

Step 3. Depress the horn button and note the voltage reading (many cars require the ignition key to be turned on before the horn will

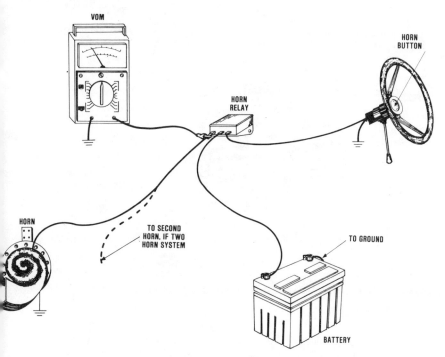

Figure 12-15: A typical electrical diagram of a horn system showing how to connect a voltmeter for circuit testing

blow). You should read a voltage that is slightly below the battery voltage, about 10 or 11 volts in a 12 volt system. If you read zero volts, there is a break in the wiring, a trouble in the horn relay, or a defective horn button.

Step 4. On the other hand, if you read a low voltage — say about 6 or 7 volts in a 12 volt system — there is a partial short or the relay points are not making good contact. A couple of checks with an ohmmeter should quickly locate the fault. However, don't forget to disconnect the battery before making resistance measurements.

12.17 AUTOMOTIVE FUSE TEST

Test Equipment:

Voltmeter

Test Setup:

See Figure 12-16.

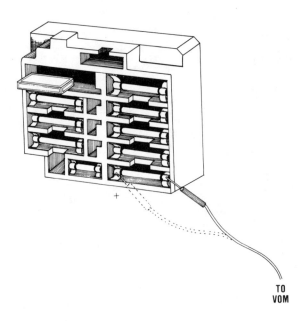

Figure 12-16: Voltmeter connections for testing a fuse in-circuit

Procedure:

Step 1: Connect the voltmeter as shown in Figure 12-16. Set the voltmeter to read slightly more than 12 VDC, with the engine *not* running.

Step 2. Turn on the ignition switch and the switch for the particular equipment the fuse protects.

Step 3. Touch the positive lead of the voltmeter to both sides of the fuse. A voltmeter reading of about 12 VDC on both sides of the fuse indicates the fuse is good. On the other hand, a reading on just one side of the fuse means the fuse is bad. Finally, a reading of zero voltage on both sides of the fuse means there is an open circuit in the wiring between the battery and fuse block.

INDEX

251